edexcel
advancing learning, changing lives

Student Book

Series Editors: *Trevor Johnson & Tony Clough*

Aim High 2
Aiming for Grade A/A*
in Edexcel GCSE Mathematics

Linear and Modular

A PEARSON COMPANY

Contents

SHAPE, SPACE AND MEASURES

Contents

Introduction

Welcome to *Aim High* – the grade-boosting revision guide for exam success.

Edexcel has studied the performance in exams of hundreds of thousands of students to find out which topics students find most difficult and the common errors they make. Written by examiners and based on Edexcel's own results data, *Aim High* targets all the areas where students regularly lose most marks in the exams. It helps you prioritise the skills and techniques on which to focus your revision, how to avoid losing marks and how to gain marks through showing your workings.

Aim High 2 is designed to help you achieve Grade A/A*. It is full of advice, hints and tips so you gain as many marks as possible from your working and answers. By focusing your revision on these skills and techniques, you improve your chance of getting the grade you need.

After this introduction is some important general exam advice and exam hints. The key maths topics are covered on pages 10 to 128. All the topics in the book have been chosen because students regularly lose marks in these areas and your teacher will help you decide which skills you should focus on. You can use the colour-coding to help you plan your revision programme: full steam ahead now with the essential green topics; get ready too for the important amber topics; and, finally, make sure your red skills don't hold you back. At the end of the book you will find a list of key terms and exam vocabulary, and answers to the questions.

Features of an *Aim High* topic

You may not need to practise all the topics in this book. Your teacher will tell you which ones to focus on. Tackle the important green topics first. Then work through the amber topics. If you still have time, work on the red topics too.

These are the maths skills you need to practise in order to succeed at the topic.

Each unit starts with a reminder of what you need to know on this topic. Make sure you can remember these facts.

Other sections in the *Aim High 2* book on this topic which you may find helpful.

Questions similar to those you may come across in the exam.

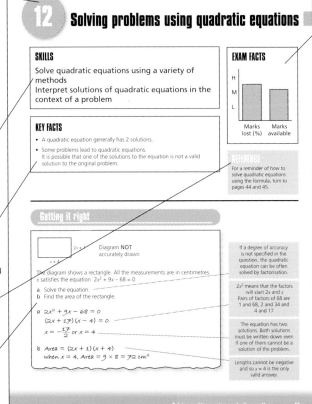

The first bar in the bar chart shows how students perform when tested: the higher the bar the worse students perform at these skills. The second bar shows how many marks are available in the exams from questions like these. In this example, quite a lot of marks are available *but* students lose a high percentage of those marks.

In this example from page 116, although the number of marks available for this topic in the exams is relatively low (right hand bar) students regularly lose a high proportion of those marks (left hand bar).

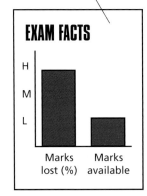

Introduction

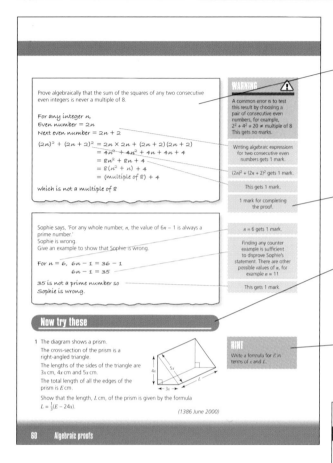

Prove algebraically that the sum of the squares of any two consecutive even integers is never a multiple of 8.

For any integer n,
Even number = $2n$
Next even number = $2n + 2$

$(2n)^2 + (2n + 2)^2 = 2n \times 2n + (2n + 2)(2n + 2)$
$= 4n^2 + 4n^2 + 4n + 4n + 4$
$= 8n^2 + 8n + 4$
$= 8(n^2 + n) + 4$
$= (\text{multiple of } 8) + 4$

which is not a multiple of 8

WARNING ⚠

A common error is to test this result by choosing a pair of consecutive even numbers, for example, $2^2 + 4^2 = 20 \neq$ multiple of 8. This gets no marks.

Writing algebraic expressions for two consecutive even numbers gets 1 mark.

$(2n)^2 + (2n + 2)^2$ gets 1 mark.

This gets 1 mark.

1 mark for completing the proof.

Sophie says, 'For any whole number, n, the value of $6n - 1$ is always a prime number.'
Sophie is wrong.
Give an example to show that Sophie is wrong.

For $n = 6$, $6n - 1 = 36 - 1$
$6n - 1 = 35$

35 is not a prime number so Sophie is wrong.

$n = 6$ gets 1 mark.

Finding any counter example is sufficient to disprove Sophie's statement. There are other possible values of n, for example $n = 11$

This gets 1 mark.

Now try these

1 The diagram shows a prism.
The cross-section of the prism is a right-angled triangle.
The lengths of the sides of the triangle are $3x$ cm, $4x$ cm and $5x$ cm.
The total length of all the edges of the prism is E cm.

Show that the length, L cm, of the prism is given by the formula
$L = \frac{1}{3}(E - 24x)$.

(1386 June 2000)

HINT

Write a formula for E in terms of x and L.

60   Algebraic proofs

You can gain marks from your workings even if your final answer is wrong. The worked examples have been written to show you how to gain maximum marks from your answers.

Advice on how to pick up marks and how to avoid losing marks.

Questions give you a chance to practise the skills. The questions in these exercises progress from easy to more difficult.

If you need help to get started, the hints tell you how to tackle the question.

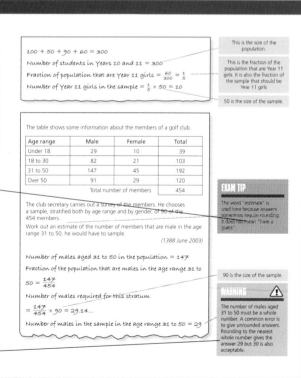

$100 + 50 + 90 + 60 = 300$

Number of students in Years 10 and 11 = 300

Fraction of population that are Year 11 girls = $\frac{60}{300} = \frac{1}{5}$

Number of Year 11 girls in the sample = $\frac{1}{5} \times 50 = 10$

This is the size of the population.

This is the fraction of the population that are Year 11 girls. It is also the fraction of the sample that should be Year 11 girls

50 is the size of the sample.

The table shows some information about the members of a golf club.

Age range	Male	Female	Total
Under 18	29	10	39
18 to 30	82	21	103
31 to 50	147	45	192
Over 50	91	29	120
Total number of members			454

The club secretary carries out a survey of the members. He chooses a sample, stratified both by age range and by gender, of 90 of the 454 members.
Work out an estimate of the number of members that are male in the age range 31 to 50, he would have to sample.

(1388 June 2003)

Number of males aged 31 to 50 in the population = 147

Fraction of the population that are males in the age range 31 to

$50 = \frac{147}{454}$

Number of males required for this stratum

$= \frac{147}{454} \times 90 = 29.14...$

Number of males in the sample in the age range 31 to 50 = 29

EXAM TIP

The word "estimate" is used here because answers sometimes require rounding. It does not mean "have a guess."

90 is the size of the sample.

WARNING ⚠

The number of males aged 31 to 50 must be a whole number. A common error is to give unrounded answers. Rounding to the nearest whole number gives the answer 29 but 30 is also acceptable.

General tips on good habits to help you stay on track.

Avoid common mistakes.

Good luck with your revision!

Exam advice

- **Make sure you have all the necessary equipment.**

  This includes a 30 cm ruler, a protractor, a pair of compasses and, when allowed, a calculator.

- **Write in black or blue ink.**

  Don't use pencil or fluorescent ink. Now that scripts are marked online, it is vital that answers can be read by a scanner.

- **For diagrams use an HB pencil, but it should not be too sharp.**

  For scanning purposes, a harder pencil or a very sharp one does not show up well on a grid.

- **Show working in the space provided for each question.**

  Don't go outside the working space allowed for each question and don't write on the formulae page or on blank pages. If necessary, ask for a supplementary answer sheet and use that.

- **Don't alter your working – cross it out and replace it.**

  If you realise you have made a mistake, cross out the error with a single line and replace it with the correct working.

- **Don't give the marker a choice of answers or methods.**

- **Don't take measurements from a diagram if you are told that it is not accurately drawn.**

  Sometimes a question specifically instructs you to find the length of a line or the size of an angle by measuring but the statement *Diagram NOT accurately drawn* alongside a diagram means that taking measurements from the diagram will not give correct answers.

- **Tracing paper is useful for transformations.**

  In the examination, you can ask for tracing paper, which may help you answer questions on reflections and rotations.

- **Make sure your calculator is set in degree mode.**

  Every year, marks are lost on trigonometry questions, because calculators are in the wrong mode (rad or grad).

  To check that your calculator is in degree mode, key in $\boxed{\text{Sin}}$ $\boxed{9}$ $\boxed{0}$ $\boxed{=}$ (or $\boxed{9}$ $\boxed{0}$ $\boxed{\text{Sin}}$ $\boxed{=}$), which should give a display of 1

Exam hints

- Show all your working.

Example	Notes

Q   Solve $5x - 7 = 2x - 5$
 	
 A $5x - 2x = 7 - 5$ ✓
        $3x = 2$ ✓
        $x = 1\frac{1}{2}$ ✗
	An incorrect answer with no working can score no marks. An incorrect answer with correct working will often receive the majority of the marks. The attempt in the example would score 2 marks out of 3 but, without the working, it would score no marks.

- Before rounding, show more figures than the question asks for.

Example	Notes
Q Find the circumference of a circle with a diameter of 9.5 cm. Give your answer correct to 1 decimal place. A $\pi \times 9.5 = 29.8451...$ ✓ Circumference = 29.9 cm ✗	29.8451… shows that the correct calculation was keyed into the calculator and that the error occurs at the rounding stage.

- Make a rough estimate of calculations.

Example	Notes

Q   Find the area of a circle with a radius of 4.2 cm. Give your answer correct to 1 decimal place.
 	
 A $\pi \times 4.2^2$ ✓
        $= 26.38......$ ✗
        Area $= 26.4 \text{ cm}^2$ ✗
        Check: A rough estimate is 3×4^2
        $3 \times 4^2 = 3 \times 16 = 48$
	The usual method for finding rough estimates is to round each number to 1 significant figure. A rough estimate does not tell you whether your answer is right but it does tell you whether it is reasonable; in this case, it is not.

- Whenever possible, ask yourself 'Is my answer sensible?'

Example	Notes

Q   Work out the value of x. Give your answer correct to 1 decimal place.
 	
      8 cm x cm   5 cm
 	
 A $x^2 = 8^2 + 5^2$ ✗
        $= 64 + 25 = 89$ ✗
        $x = \sqrt{89} = 9.4$ ✗
	This answer is **not** sensible as, in a right-angled triangle, the longest side is the hypotenuse, the one opposite the right angle.
	 (The first line of working should be $8^2 = x^2 + 5^2$)

- Whenever possible, check your answers.

Example	Notes

Q $v = u + at$
 Find a when $v = 4$, $u = 10$ and $t = 2$
 	
 A $4 \ = 10 + 2a$ ✓
       $2a = 4 - 10 = -6$ ✓
       $a \ = -3$ ✓
 Check by substituting the values of u, a, and t into $v = u + at$ to find the value of v.
       $10 - 3 \times 2 = 10 - 6 = 4$ ✓
	There are many ways of checking answers. When solving equations, for example, check that your solution fits the original equation. It is sometimes possible, as a check, to work a calculation backwards.

1 Reverse percentages

SKILL

Work out the original quantity given the final value after a percentage increase or a percentage decrease

EXAM FACTS

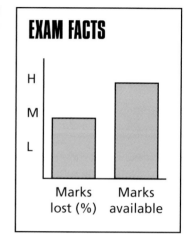

Marks lost (%)   Marks available

KEY FACTS

Method 1 – using multipliers

- work out the original multiplier,
- divide the final value by the original multiplier,

  (To find the original multiplier:

  - for a percentage increase, add the given percentage to 100% then convert this percentage to a decimal.

  - for a percentage decrease, subtract the given percentage from 100% then convert this percentage to a decimal).

Method 2 – using percentages

- find the percentage of the original quantity that is represented by the final value (For a percentage increase, add the given percentage to 100%; for a percentage decrease, subtract the given percentage from 100%),
- divide the final value by this percentage to find 1% of the original quantity,
- multiply by 100 to find the original quantity.

Getting it right

In a sale all the prices are reduced by 30%.
The sale price of a jacket is £28.
Work out the price of the jacket before the sale.

(1385 November 2000)

Method 1 – using multipliers

$100\% - 30\% = 70\%$

$70\% = \frac{70}{100} = 0.7$

WARNING

The most common error is to reduce £28 by 30%.

Find the multiplier for a **decrease** of 30% and write the multiplier as a decimal.

Let the original price be x

$x \times 0.7 = 28$

$x = \dfrac{28}{0.7}$

$= 40$

The price of the jacket before the sale was £40

Method 2 — using percentages

$100\% - 30\% = 70\%$

1% of the original price is $\dfrac{28}{70}$

100% of the original price is $\dfrac{28}{70} \times 100 = 40$

The price of the jacket before the sale was £40

> The original price was multiplied by 0.7 to give £28 Write this as an equation.

> Solve the equation.

> (Check 40 × 0.7 = 28)

> £28 represents 70% of the original price.

> Divide 28 by 70 to find 1% of the original price.

> The original price is 100% so multiply 1% of the original price by 100

> (Check 70% of 40 = 28)

The selling price of a computer is the **list price** plus VAT at $17\frac{1}{2}\%$.
A computer has a selling price of £1292.50
Work out the **list price** of this computer.

(1385 November 2001)

Method 1 — using multipliers

$100\% + 17\frac{1}{2}\% = 117.5\%$

$117.5\% = \dfrac{117.5}{100} = 1.175$

Let the original price be x

$x \times 1.175 = 1292.50$

$x = \dfrac{1292.50}{1.175}$

$= 1100$

The list price of this computer is £1100

Method 2 — using percentages

$100\% + 17\frac{1}{2}\% = 117.5\%$

1% of the original price is $\dfrac{1292.50}{117.5}$

100% of the original price is $\dfrac{1292.50}{117.5} \times 100 = 1100$

The list price of this computer is £1100

WARNING

The most common error is to increase £1292.50 by $17\frac{1}{2}\%$.

> Find the multiplier for an **increase** of 17.5% and write the multiplier as a decimal.

> The original amount was multiplied by 1.175 to give £1292.50 Write this as an equation.

> Solve the equation.

> (Check 1100 × 1.175 = 1292.5)

> £1292.50 represents 117.5% of the original price.

> Divide 1292.50 by 117.5 to find 1% of the original price.

> The original price is 100% so multiply 1% of the original price by 100

> (Check 117.5% of 1100 = 1292.5)

1 In a sale, all the normal prices are reduced by 15%.
   The sale price of a suit is £110.50
   Work out the normal price of the suit.

2 During the last year the number of pupils at a school increased by 8%
   to 1944
   Work out the number of pupils in the school before the increase.

3 The price of a new DVD recorder is £199.75
   This price includes Value Added Tax (VAT) at 17½%.
   Work out the cost of the DVD recorder *before* VAT was added.

4 Seth invests some money in a bank account.
   Interest is paid at a rate of 4.3% per annum.
   After 1 year, there is £354.62 in his account.
   How much money did Seth invest?

5 Loft insulation reduces annual heating costs by 20%.
   After he insulated his loft, Curtley's annual heating cost was £520
   Work out what Curtley's annual heating cost would have been, if he had
   not insulated his loft.

   (1387 November 2003)

6 In a sale, normal prices are reduced by 20%.
   Andrew bought a saddle for his horse in the sale.
   The sale price of the saddle was £220
   Calculate the normal price of the saddle.

   **SALE
   20% OFF**

   (1387 June 2005)

7 In a sale, normal prices are reduced by 14%.
   The sale price of a digital camera is £129.86
   Work out the normal price of the digital camera.

   (1387 November 2005)

8 In a sale, normal prices are reduced by 12%.
   The sale price of a DVD player is £242
   Work out the normal price of the DVD player.

   (1388 March 2003)

9 A garage sells cars.
   It offers a discount of 20% off the normal price for cash.
   Dave pays £5200 cash for a car.
   Calculate the normal price of the car.

   (1388 June 2003)

10 The price of all rail season tickets to London increased by 4%.
   After the increase, the price of a rail season ticket from Brighton to London
   was £2828.80
   Work out the price before this increase.

   (1387 November 2006)

2 Standard form

SKILLS

Write an ordinary number as a number in standard form

Write a number in standard form as an ordinary number

Calculate with numbers in standard form

EXAM FACTS

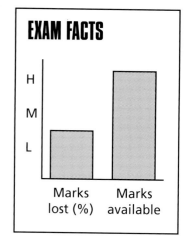

Marks lost (%) Marks available

KEY FACTS

- A number written in standard form has the form $a \times 10^n$ where $1 \leqslant a < 10$ and n is an integer.

- To enter a number in standard form into a calculator use the ⌐^⌐ or ⌐EXP⌐ button.

  For example, to enter 6.3×10^4 key in either

  ⌐6⌐ ⌐.⌐ ⌐3⌐ ⌐×⌐ ⌐1⌐ ⌐0⌐ ⌐^⌐ ⌐4⌐ or ⌐6⌐ ⌐.⌐ ⌐3⌐ ⌐EXP⌐ ⌐4⌐

- Some of the newer calculators have the key ⌐×10^x⌐ instead of the ⌐EXP⌐ key.

Getting it right

i   Write 638 000 in standard form.

ii Write 5.03×10^{-2} as an ordinary number.

(1388 March 2005)

i $638000 = 6.38 \times 10^5$

ii $5.03 \times 10^{-2} = 0.0503$

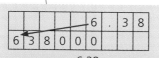

$a = 6.38$
The digits have moved from 5 places to the left so $n = 5$

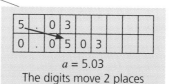

$a = 5.03$
The digits move 2 places to the right.

WARNING ⚠

1 Remember to include all the digits in the given number when writing it in standard form. A common error is to omit the 3 and the 8 and just write 6×10^5 for the answer.

2 A common error is in choosing the value for a. For example, the value of 638×10^3 also equals 638000 but 638×10^3 is not in standard form as $a = 638$ does not satisfy the condition $1 \leqslant a < 10$

A spaceship travelled for 6×10^2 hours at a speed of 8×10^4 km/h.

a   Calculate the distance travelled by the spaceship.
    Give your answer in standard form.

One month an aircraft travelled 2×10^5 km.
The next month the aircraft travelled 3×10^4 km.

b   Calculate the total distance travelled by the aircraft in the two months.
    Give your answer as an ordinary number.

(1387 June 2003)

a   Distance $= 6 \times 10^2 \times 8 \times 10^4$

    $= 6 \times 8 \times 10^2 \times 10^4$

    $= 48 \times 10^{2+4}$

    $= 48 \times 10^6$

    $= 4.8 \times 10^1 \times 10^6$

    $= 4.8 \times 10^7$

The spaceship travelled 4.8×10^7 km.

b   Distance $= 2 \times 10^5 + 3 \times 10^4$

    $= 200\,000 + 30\,000$

    $= 230\,000$ km

The aircraft travelled $230\,000$ km.

a   Work out the value of $\dfrac{3 \times 10^{-6}}{4 \times 10^{-4}}$
    Give your answer in standard form.

The distance from Earth to the star Proxima Centauri is 4.22 light years.
1 light year $= 9.641 \times 10^{12}$ km.

b   Work out the distance from Earth to the star Proxima Centauri.
    Give your answer in standard form correct to 3 significant figures.
    (1385 November 2001)

a $\dfrac{3 \times 10^{-6}}{4 \times 10^{-4}} = 7.5 \times 10^{-3}$

b   Distance $= 4.22 \times 9.641 \times 10^{12}$

    $= 4.068502 \times 10^{13}$

    $= 4.07 \times 10^{13}$ km (to 3 s.f.)

Now try these

1 **a** Write 47 500 000 in standard form.
   b Write 0.000 06 in standard form.

(1388 January 2003)

2 **a** Write the number 50 000 000 in standard form.
   b Write the number 0.000 082 in standard form.

(1388 January 2004)

3 **a** Write 3.8×10^3 as an ordinary number.
   b Write the number 0.000 45 in standard form.

(1388 March 2004)

4 **a** Write 65 200 in standard form.
   b Write 8.36×10^{-2} as an ordinary number.

(1388 November 2005)

5 **a** Write 0.032 in standard form.
   b Write 1.58×10^4 as an ordinary number.

(1388 November 2005)

6 **a** Write 76 800 000 in standard form.
   b Write 0.000 35 in standard form.

(1388 March 2006)

7 Write in standard form
   a 456 000
   b 0.000 34
   c 16×10^7

(1387 June 2006)

8 **a** **i** Write the number 5.01×10^4 as an ordinary number.
         ii Write the number 0.0009 in standard form.
   b Multiply 4×10^3 by 6×10^5.
      Give your answer in standard form.

(1385 June 2001)

9 **a** **i** Write 40 000 000 in standard form.
         ii Write 3×10^{-5} as an ordinary number.
   b Work out the value of $3 \times 10^{-5} \times 40\,000\,000$
      Give your answer in standard form.

(1387 November 2004)

10 Work out $(3.4 \times 10^{12}) \div (1.2 \times 10^{-3})$
   Give your answer in standard form, correct to 3 significant figures.

(1388 March 2003)

11 Calculate the value of $\dfrac{5.98 \times 10^8 + 4.32 \times 10^9}{6.14 \times 10^{-2}}$

Give your answer in standard form, correct to 3 significant figures.

(1385 June 2001)

12 a Write 5 720 000 in standard form.

$p = 5\,720\,000$
$q = 4.5 \times 10^5$

  b Find the value of $\dfrac{p - q}{(p + q)^2}$

Give your answer in standard form, correct to 2 significant figures.

(1388 November 2005)

13 $x = \sqrt{\dfrac{p + q}{pq}}$

$p = 4 \times 10^8$
$q = 3 \times 10^6$

Find the value of x.

Give your answer in standard form, correct to 2 significant figures.

(1388 March 2005)

14 Work out $\dfrac{4.07 \times 10^3 \times 2.17 \times 10^5}{5.1 \times 10^{-4}}$

Give your answer in standard form, correct to 2 significant figures.

(1385 November 2002)

15 When you are h feet above sea level, you can see d miles to the horizon,

where $d = \sqrt{\dfrac{3h}{2}}$

Calculate the value of d when $h = 8.4 \times 10^3$

Give your answer in standard form correct to 3 significant figures.

(1387 November 2006)

16 A nanosecond is 0.000 000 001 second.
  a Write the number 0.000 000 001 in standard form.

  A computer does a calculation in 5 nanoseconds.
  b How many of these calculations can the computer do in 1 second?
     Give your answer in standard form.

(1387 June 2004)

17 420 000 carrot seeds weigh 1 gram.
  Each carrot seed weighs the same.
  a Write the number 420 000 in standard form.
  b Calculate the weight, in grams, of one carrot seed.
     Give your answer in standard form, correct to 2 significant figures.

(1385 June 2002)

3 **Bounds**

SKILL

Use bounds in calculations

EXAM FACTS

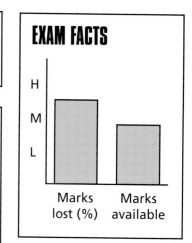

KEY FACTS

- Measurements given to the nearest whole unit may be inaccurate by up to one half of a unit below and one half of a unit above the measurement.

  For example
  4 cm correct to the nearest centimetre
    Maximum possible length = 4.5 cm
    Minimum possible length = 3.5 cm

  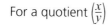

- The true value of a number given to a number of significant figures or decimal places lies between its upper bound and its lower bound.

  For example
  6.8 correct to 1 decimal place
    Upper bound = 6.85
    Lower bound = 6.75

  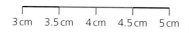

  320 correct to 2 significant figures
    Upper bound = 325
    Lower bound = 315

- When carrying out calculations with numbers that have been rounded, it is possible to calculate the range of answers that can be produced. Whether the upper bound$_{(ub)}$ or lower bound$_{(lb)}$ is used depends on the type of calculation.

  For a sum $(x + y)$
  $(x + y)_{ub} = x_{ub} + y_{ub}$
  $(x + y)_{lb} = x_{lb} + y_{lb}$

  For a product (xy)
  $(xy)_{ub} = x_{ub} \times y_{ub}$
  $(xy)_{lb} = x_{lb} \times y_{lb}$

  For a difference $(x - y)$
  $(x - y)_{ub} = x_{ub} - y_{lb}$
  $(x - y)_{lb} = x_{lb} - y_{ub}$

  For a quotient $\left(\frac{x}{y}\right)$
  $\left(\frac{x}{y}\right)_{ub} = \left(\frac{x_{ub}}{y_{lb}}\right)$

  $\left(\frac{x}{y}\right)_{lb} = \left(\frac{x_{lb}}{y_{ub}}\right)$

A common error would be to use the given values rather than the upper bounds to find the perimeter.

A field is in the shape of a rectangle.
The length of the field is 340 m, to the nearest metre.

339 339.5 340 340.5 341

Numbers in the red interval are closer to 340 than to 339 or 341

The width of the field is 117 m, to the nearest metre.

116 116.5 117 117.5 118

Numbers in the red interval are closer to 117 than to 116 or 118

Calculate the upper bound for the perimeter of the field.

(1388 June 2003)

Perimeter = 2 × (length + width)

The length and width of the rectangle are added to find the perimeter so use the upper bound of each measurement.

Upper bound of length = 340.5 m
Upper bound of width = 117.5 m

Write down both the upper bounds.

Upper bound for perimeter = 2 × (340.5 + 117.5)
= 916 m

2 × (Upper bound of length + Upper bound of width)

Elliot did an experiment to find the value of g m/s^2, the acceleration due to gravity. He measured the time, T seconds, that a block took to slide L m down a smooth slope of angle $x°$.

L m

$x°$

He then used the formula

$$g = \frac{2L}{T^2 \sin x°}$$

to calculate an estimate for g.

$T = 1.3$ correct to 1 decimal place.

Numbers in the red interval are closer to 1.3 than to 1.2 or 1.4

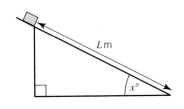

1.2 1.25 1.3 1.35 1.4

$L = 4.50$ correct to 2 decimal places.

4.49 4.495 4.50 4.505 4.51

Numbers in the red interval are closer to 4.50 than to 4.49 or 4.51

$x = 30$ correct to the nearest integer.

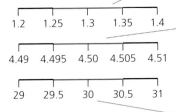

29 29.5 30 30.5 31

Calculate the lower bound and the upper bound for the value of g.
Give your answers correct to 3 decimal places.

(1387 June 2003)

Numbers in the red interval are closer to 30 than to 29 or 31

Upper bound of $T = 1.35$
Lower bound of $T = 1.25$

Upper bound of $L = 4.505$
Lower bound of $L = 4.495$

Upper bound of $x = 30.5$
Lower bound of $x = 29.5$

Lower bound of $g = \dfrac{2 \times 4.495}{1.35^2 \times \sin 30.5°}$

$= 9.71903...$

$= 9.719$ (to 3 d.p.)

Upper bound of $g = \dfrac{2 \times 4.505}{1.25^2 \times \sin 29.5°}$

$= 11.71024...$

$= 11.710$ (to 3 d.p.)

WARNING

A common error is to give the upper bound and lower bound of 30° (correct to the nearest integer) as 35° and 25° respectively.

$\dfrac{2 \times \text{Lower bound of } L}{(\text{Upper bound of } T)^2 \times \sin (\text{Upper bound of } x)}$

EXAM TIP

Write down at least 4 decimal places of the calculator display.

An answer correct to 3 decimal places is required.

$\dfrac{2 \times \text{Upper bound of } L}{(\text{Lower bound of } T)^2 \times \sin (\text{Lower bound of } x)}$

EXAM TIP

Write down at least 4 decimal places of the calculator display, before rounding your answer.

Now try these

1 To the nearest integer, $x = 8$, $y = 5$
  Work out the upper bounds of
  a xy b $x + y$ c $x - y$ d $\dfrac{x}{y}$

2 To the nearest integer, $x = 12$, $y = 9$
  Work out the lower bounds of
  a xy b $x + y$ c $x - y$ d $\dfrac{x}{y}$

3 A garden is in the shape of a rectangle.
  The length of the garden is 24 m, correct to the nearest metre.
  The width of the garden is 10 m, correct to the nearest metre.
  a Calculate the lower bound for the area of the garden.
  b Calculate the upper bound for the area of the garden.

4 $f = \dfrac{n}{r}$

  $n = 1.49$ correct to 2 decimal places.
  $r = 0.5$ correct to 1 significant figure.
  Write down the value of n and the value of r which are substituted into the formula to calculate the upper bound of f.

(1388 November 2006)

5 To the nearest centimetre, $x = 4$ cm and $y = 6$ cm.

   a Calculate the upper bound for the value of xy.

   b Calculate the lower bound for the value of $\frac{x}{y}$.

     Give your answer correct to 3 significant figures.

(1385 June 2002)

6

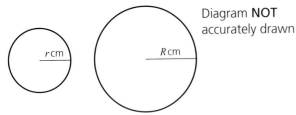

Diagram **NOT**
accurately drawn

The diagram represents two metal spheres of different sizes.
The radius of the smaller sphere is r cm.
The radius of the larger sphere is R cm.
$r = 1.7$ correct to 1 decimal place.
$R = 31.0$ correct to 3 significant figures.

   a Write down the upper and lower bounds of r and R.

   b Find the smallest possible value of $R - r$.

The larger sphere of radius R cm was melted down and used to make
smaller spheres of radius r cm.

   c Calculate the smallest possible number of spheres that could be made.

(1387 November 2003)

7 Martin won the 400 metre race in the school sports with a time of
1 minute.
The distance was correct to the nearest centimetre.
The time was correct to the nearest tenth of a second.

Work out the upper bound and the lower bound of Martin's speed in km/h.
Give your answers correct to 5 significant figures.

(1387 June 2004)

8 The length of a rectangle is 6.7 cm, correct to 2 significant figures.

   a For the length of the rectangle write down

     i the upper bound,

     ii the lower bound.

The area of the rectangle is 26.9 cm², correct to 3 significant figures.

   b **i** Calculate the upper bound for the width of the rectangle.
       Write down all the figures on your calculator display.

     ii Calculate the lower bound for the width of the rectangle.
       Write down all the figures on your calculator display.

(1387 June 2006)

9 A clay bowl is in the shape of a hollow hemisphere.

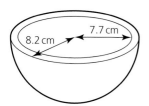

Diagram **NOT**
accurately drawn

The external radius of the bowl is 8.2 cm.
The internal radius of the bowl is 7.7 cm.
Both measurements are correct to the nearest 0.1 cm.
The upper bound for the volume of clay is $k\pi$ cm^3.
Find the exact value of k.

(1388 March 2003)

10

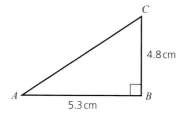

Diagram **NOT**
accurately drawn

In triangle ABC, angle $ABC = 90°$.
$AB = 5.3$ cm, correct to 2 significant figures.
$BC = 4.8$ cm, correct to 2 significant figures.
The base, AB, of the triangle is horizontal.

i   Calculate the lower bound for the gradient of the line AC.
ii   Calculate the upper bound for the gradient of the line AC.

(1387 November 2006)

4 Surds

SKILLS

Use surds in calculations
Simplify surds
Rationalise surds

EXAM FACTS

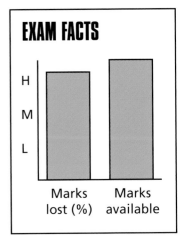

KEY FACTS

- A surd is a root of a number that does not have an exact value.

- To simplify surds, use the facts that

  $\sqrt{m} \times \sqrt{n} = \sqrt{mn}$

  $\dfrac{\sqrt{m}}{\sqrt{n}} = \sqrt{\dfrac{m}{n}}$

  $\sqrt{m} \times \sqrt{m} = m$

- To rationalise the denominator of $\dfrac{a}{\sqrt{b}}$ multiply the fraction by $\dfrac{\sqrt{b}}{\sqrt{b}}$.

Getting it right

Find the exact value of $(\sqrt{7} + \sqrt{5})(\sqrt{7} - \sqrt{5})$

(1385 November 2001)

Expand the brackets.

$(\sqrt{7} + \sqrt{5})(\sqrt{7} - \sqrt{5})$

$= (\sqrt{7} \times \sqrt{7}) - (\sqrt{7} \times \sqrt{5}) + (\sqrt{5} \times \sqrt{7}) - (\sqrt{5} \times \sqrt{5})$

$= 7 - 5$

$= 2$

WARNING

A common error is to forget to simplify $\sqrt{7} \times \sqrt{7}$ and $\sqrt{5} \times \sqrt{5}$

$\sqrt{7} \times \sqrt{7} = 7$ and
$\sqrt{5} \times \sqrt{5} = 5$

Write $\dfrac{\sqrt{18} + 10}{\sqrt{2}}$ in the form $p + q\sqrt{2}$, where p and q are integers.

(1388 June 2005)

$$\dfrac{\sqrt{18} + 10}{\sqrt{2}} = \dfrac{\sqrt{18} + 10}{\sqrt{2}} \times \dfrac{\sqrt{2}}{\sqrt{2}}$$

$$= \dfrac{\sqrt{18} \times \sqrt{2} + 10\sqrt{2}}{\sqrt{2} \times \sqrt{2}}$$

$$= \dfrac{\sqrt{36} + 10\sqrt{2}}{2}$$

$$= \dfrac{6 + 10\sqrt{2}}{2}$$

$$= \dfrac{6}{2} + \dfrac{10\sqrt{2}}{2}$$

$$= 3 + 5\sqrt{2}$$

Multiply by $\dfrac{\sqrt{2}}{\sqrt{2}}$

WARNING

A common error is to write $10\sqrt{2}$ as $\sqrt{20}$

Use $\sqrt{m} \times \sqrt{n} = \sqrt{m \times n}$ to write $\sqrt{18} \times \sqrt{2}$ as $\sqrt{36}$

$\sqrt{2} \times \sqrt{2} = 2$

Divide both 6 and $10\sqrt{2}$ by 2

Now try these

1 Find the value of k.

  a $\sqrt{12} = k\sqrt{3}$
  b $\sqrt{75} = k\sqrt{3}$
  c $\sqrt{20} = k\sqrt{5}$
  d $\sqrt{200} = k\sqrt{2}$
  e $\sqrt{162} = k\sqrt{2}$
  f $\sqrt{63} = k\sqrt{7}$

2 Expand these expressions. Give your answers in the form $a + b\sqrt{c}$ where a, b and c are integers.

  a $(2 + \sqrt{3})(5 - \sqrt{3})$
  b $(\sqrt{2} + 7)(1 - \sqrt{2})$
  c $(3 + \sqrt{5})^2$
  d $(4 + \sqrt{3})(2 - \sqrt{3})$

3 Rationalise the denominators and simplify your answers.

  a $\dfrac{1}{\sqrt{3}}$
  b $\dfrac{3}{\sqrt{5}}$
  c $\dfrac{4}{\sqrt{2}}$
  d $\dfrac{15}{\sqrt{5}}$

4 Write the following in the form $p + q\sqrt{3}$, where p and q are integers.

  a $\dfrac{6 + \sqrt{3}}{\sqrt{3}}$
  b $\dfrac{15 + \sqrt{12}}{\sqrt{3}}$

5 The length of a rectangle is $(2 + \sqrt{3})$ cm.

The width of the rectangle is $(5 - \sqrt{3})$ cm.

    a Work out the perimeter of the rectangle.

    b Work out the area of the rectangle, give your answer in the form
$p + q\sqrt{3}$, where p and q are integers.

6 Write $\dfrac{6 + \sqrt{8}}{\sqrt{2}}$ in the form $p + q\sqrt{2}$ where p and q are integers.

7 Write $\dfrac{9 - \sqrt{12}}{\sqrt{3}}$ in the form $a + b\sqrt{3}$ where a and b are integers.

8 **a** Write $\sqrt{8}$ in the form $m\sqrt{2}$, where m is an integer.

    b Write $\sqrt{50}$ in the form $k\sqrt{2}$, where k is an integer.

    c Rationalise $\dfrac{1 + \sqrt{2}}{\sqrt{2}}$

(1387 November 2006)

9 $8\sqrt{8}$ can be written in the form 8^k

    a Find the value of k.

$8\sqrt{8}$ can also be expressed in the form $m\sqrt{2}$ where m is a positive integer.

    b Express $8\sqrt{8}$ in the form $m\sqrt{2}$

    c Rationalise the denominator of $\dfrac{1}{8\sqrt{8}}$

Give your answer in the form $\dfrac{\sqrt{2}}{p}$ where p is a positive integer.

(1387 June 2006)

10 **a** Rationalise $\dfrac{1}{\sqrt{7}}$

    b **i** Expand and simplify $(\sqrt{3} + \sqrt{15})^2$

Give your answer in the form $n + m\sqrt{5}$, where n and m are integers.

      ii

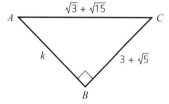

Diagram **NOT** accurately drawn

All measurements on the triangle are in centimetres.
ABC is a right-angled triangle.
k is a positive integer.
Find the value of k.

(1387 November 2005)

5 Solving equations that have fractions

EXAM FACTS

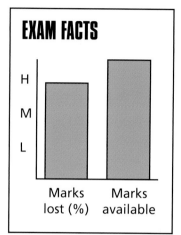

Getting it right

Solve the equation $2(x + 4) = \frac{x}{2} - 1$

(1388 January 2002)

$2(x + 4) = \frac{x}{2} - 1$

$2x + 8 = \frac{x}{2} - 1$

$2(2x + 8) = 2\left(\frac{x}{2} - 1\right)$

$4x + 16 = x - 2$

$4x - x = -2 - 16$

$3x = -18$

$x = -6$

WARNING

A common error is to use trial and improvement methods.

$2x + 8$ gets 1 mark.

Multiplying both sides by 2 gets 1 mark.

WARNING

A common error is to forget to multiply the -1 by 2

1 mark for the correct answer.

Collecting the x terms on one side and the numbers on the other (with correct signs) gets 1 mark.

Solve the equation $\dfrac{7}{x+1} + \dfrac{1}{x-2} = 4$

$$\dfrac{7}{x+1} + \dfrac{1}{x-2} = 4$$

$$(x+1)(x-2)\dfrac{7}{(x+1)} + (x+1)(x-2)\dfrac{1}{(x-2)} = 4(x+1)(x-2)$$

$$\cancel{(x+1)}(x-2)\dfrac{7}{\cancel{(x+1)}} + (x+1)\cancel{(x-2)}\dfrac{1}{\cancel{(x-2)}} = 4(x+1)(x-2)$$

$$7(x-2) + (x+1) = 4(x+1)(x-2)$$

$$7x - 14 + x + 1 = 4(x^2 + x - 2x - 2)$$

$$8x - 13 = 4x^2 - 4x - 8$$

$$4x^2 - 12x + 5 = 0$$

$$(2x - 5)(2x - 1) = 0$$

$$x = 2.5 \text{ or } x = 0.5$$

- Showing $(x+1)(x-2)$ gets 1 mark.
- Multiplying both sides by $(x+1)(x-2)$ gets 1 mark.
- $7(x-2) + (x+1)$ gets 1 mark.
- Expanding the brackets and forming a quadratic equation gets 1 mark.
- Rearranging into this form gets 1 mark.
- Correct factorisation of $4x^2 - 12x + 5$ gets 1 mark.
- 1 mark for the correct solutions.

Now try these

1 Solve $\dfrac{5x+4}{3} = 2$ *(4400 November 2005)*

2 Solve $\dfrac{x-3}{5} = x - 5$ *(1387 June 2005)*

3 Solve $\dfrac{40-x}{3} = 4 + x$ *(1387 June 2004)*

4 Solve $\dfrac{4}{y} + 7 = 2$ *(1385 June 1998)*

5 Solve $\dfrac{x-1}{2} + \dfrac{2x+3}{4} = 1$ *(4400 May 2005)*

6 Solve $\dfrac{2}{x} + \dfrac{3}{2x} = \dfrac{1}{3}$ *(1385 November 2000)*

7 Solve $\dfrac{2}{x+1} + \dfrac{3}{x-1} = \dfrac{5}{x^2 - 1}$

8 a Solve $\dfrac{3}{x} + \dfrac{3}{2x} = 2$

   b Using your answer to part **a**, or otherwise, solve $\dfrac{3}{(y-1)^2} + \dfrac{3}{2(y-1)^2} = 2$

                                                      (1387 June 2006)

9 Solve $\dfrac{1}{x-2} + \dfrac{2}{x+4} = \dfrac{1}{3}$ *(1385 June 2001)*

10 Find the values of x, correct to 2 decimal places, that satisfy the equation

$$\dfrac{2}{(x+2)} - \dfrac{1}{(x+3)} = \dfrac{1}{2}$$

SKILL

Change the subject of a formula

KEY FACTS

- To change the subject of a formula, rearrange the terms in the formula using inverse operations.
- To change the subject of a formula where the required subject occurs twice, first collect the terms in the required subject, then factorise by taking out the required subject and rearrange the terms in the formula.

EXAM FACTS

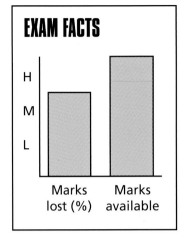

| | Marks lost (%) | Marks available |

Getting it right

Make r the subject of the formula $D = 2\pi r + 3r + 6a$

$D = 2\pi r + 3r + 6a$

$D - 6a = 2\pi r + 3r$

$D - 6a = r(2\pi + 3)$

$r = \dfrac{D - 6a}{(2\pi + 3)}$

Make R the subject of the formula $A = \pi(R + r)(R - r)$

(4400 May 2006)

$A = \pi(R + r)(R - r)$

$A = \pi(R^2 - Rr + Rr - r^2)$

$A = \pi(R^2 - r^2)$

$A = \pi R^2 - \pi r^2$

$A + \pi r^2 = \pi R^2$

$R^2 = \dfrac{A - \pi r^2}{\pi}$

$R = \sqrt{\dfrac{A - \pi r^2}{\pi}}$

Expand $(R + r)(R - r)$ by multiplying out or by recognising the difference of two squares to get $R^2 - r^2$

WARNING

A common error is to rearrange the formula so that the required subject is on the left hand side, but leaving a term including the required subject on the right hand side. This gets no marks.

Subtract $6a$ from both sides. This gets 1 mark.

Factorise the terms in r. This gets 1 mark.

Divide both sides by $(2\pi + 3)$ to get $r = \dots$

Expand the bracket and isolate the term in R. This gets 1 mark.

Divide both sides by π

The inverse of squaring is taking the square root. So take the square root of both sides to get $R = \dots$

1 Make v the subject of the formula $m(v - u) = I$

(4400 November 2004)

2 Make u the subject of the formula $v^2 = u^2 + 2as$

(1384 June 1996)

3 Make W the subject of the formula $h = \sqrt{\dfrac{W}{I}}$

(4400 May 2005)

4 Make p the subject of the formula $M = 2(a - p^2)$

5 $V = \frac{1}{3}\pi(a^2 + b^2)$

Rearrange the formula to make b the subject.

6 Make g the subject of the formula $T = 2\pi\sqrt{\dfrac{l}{g}}$

7 $f = \dfrac{uv}{u + v}$

Rearrange the formula to make v the subject.

8 $P = \dfrac{n^2 + a}{n + a}$

Rearrange the formula to make a the subject.

(1387 June 2006)

9 Rearrange the formula $y = \dfrac{k}{(x + a)^2}$ to express x in terms of k, y and a.

(1384 June 1998)

10 $\dfrac{x}{x + c} = \dfrac{p}{q}$

Make x the subject of the formula.

(1387 November 2006)

Using $y = mx + c$

SKILLS

Use $y = mx + c$ to write down the equations of lines parallel to a given line

Find the equations of straight lines perpendicular to a given line

EXAM FACTS

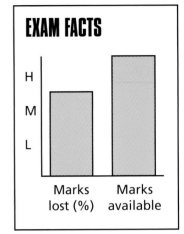

| | |
| Marks lost (%) | Marks available |

KEY FACTS

- The equation of a straight line written in the form $y = mx + c$, where m and c are numbers, has y-intercept c and gradient m.

- For a line with gradient m
    - all lines parallel to it have gradient m
    - all lines perpendicular to it have gradient $-\frac{1}{m}$.

Getting it right

A straight line, L, has equation $2y = 5 - 4x$

a   Find the gradient of the line L.

The line M is parallel to L and passes through the point (2, 1).

b   Find the equation of the line M.

a $2y = 5 - 4x$

so $y = \frac{5 - 4x}{2}$ or $y = -2x + 2.5$

The gradient of $y = -2x + 2.5$ is -2

so the gradient of the line L is -2

b   Gradient of line M is -2

so line M has an equation in the form $y = -2x + c$

$(2, 1)$ is a point on $y = -2x + c$

so $1 = -2 \times 2 + c$

so $c = 5$

The equation of the line M is $y = -2x + 5$

Write $2y = 5 - 4x$ in the form $y = mx + c$. This gets 1 mark.

Compare with $y = mx + c$ and read off the value of m for the gradient.

Line M is parallel to line L and parallel lines have equal gradients. $m = -2$ gets 1 mark.

Use $y = mx + c$

Substitute $x = 2$, $y = 1$ into $y = -2x + c$ to find c. This gets 1 mark.

Substitute $c = 5$ in $y = -2x + c$

ABCD is a rectangle.
A is the point (8, 2)
B is the point (2, 5)

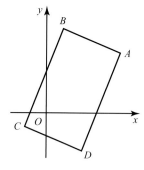

Diagram **NOT**
accurately drawn

a Find an equation of the straight line through A and B.
b Find an equation of the straight line through B and C.

a   B (2, 5)

Gradient of AB $m = -\frac{3}{6} = -\frac{1}{2}$

$y = -\frac{1}{2}x + c$

(2, 5) is a point on $y = -\frac{1}{2}x + c$

so $5 = -\frac{1}{2} \times 2 + c$

$5 = -1 + c$

$6 = c$

An equation of the line through A and B is $y = -\frac{1}{2}x + 6$

b BC is perpendicular to AB

Gradient of AB $= -\frac{1}{2}$

So gradient of BC $= -\frac{1}{-\frac{1}{2}} = 2$

BC has an equation in the form $y = 2x + c$

(2, 5) is a point on $y = 2x + c$

so $5 = 2 \times 2 + c$

$c = 1$

An equation of the line through B and C is $y = 2x + 1$

Find the gradient of the line joining (8, 2) and (2, 5). This gets 1 mark.

The line AB slopes \ so has negative gradient.

Use $y = mx + c$

Substitute $x = 2$, $y = 5$ into $y = -\frac{1}{2}x + c$ to find c. This gets 1 mark.

Substitute $c = 6$ in $y = -\frac{1}{2}x + c$

ABCD is a rectangle.

Use the fact that the gradient of a perpendicular line is $-\frac{1}{m}$. This gets 1 mark.

WARNING

A common error is to use $\frac{1}{m}$ instead of $-\frac{1}{m}$ for the gradient of a perpendicular line.

Use $y = mx + c$

Substitute $x = 2$, $y = 5$ into $y = 2x + c$ to find c. This gets 1 mark.

1 The equations of five straight lines are

$y = x - 2$, $y = 2x + 3$, $y = 3x + 2$, $y = 5x + 2$, $y = 3x - 3$

Two of the lines are parallel.
Write down the equations of these two lines.

(1384 November 1994)

2 The equations of five straight lines are

$y = 2x - 1$, $y = -2x + 3$, $y = 2x + 1$, $2y = x + 4$, $2y = 5x - 1$

Two of the lines are perpendicular.
Write down the equations of these two lines.

3 In the diagram

A is the point (0, −2),
B is the point (−4, 2),
C is the point (0, 3).

Find an equation of the line
that passes through *C* and
is parallel to *AB*.

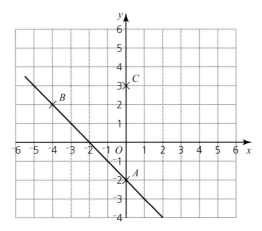

(1387 June 2006)

4 The straight line, **L**, passes
through the points (0, −1)
and (2, 3)

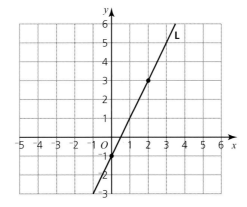

a   Work out the gradient of **L**.
b   Write down the equation of **L**.
c   Write down the equation of another line that is parallel to **L**.

(4400 November 2004)

5

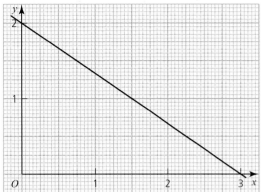

The line with equation $3y = -2x + 6$ has been drawn on the grid.
A line is drawn parallel to $3y = -2x + 6$ through the point (2, 1).
Find the equation of this line.

(1386 June 1998)

6

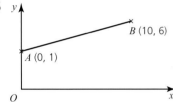

A is the point (0, 1)
B is the point (10, 6)

The equation of the straight line through *A* and *B* is $y = \frac{1}{2}x + 1$

a   Write down the equation of another straight line that is parallel
to $y = \frac{1}{2}x + 1$
b   Write down the equation of another straight line that passes through
the point (0, 1).
c   Find the equation of the line perpendicular to *AB* passing through *B*.

(1387 November 2006)

7 The straight line **L** passes through the points (−1, 2) and (2, 0)
a   Find an equation of the straight line **L**.

The straight line **M** is perpendicular to the straight line **L**.
The straight line **M** passes through the point (2, 0)
b   Find an equation of the straight line **M**.

Solving inequalities graphically

<table>
<tr><td>

SKILL

Solve inequalities by shading regions on graphs

</td></tr>
</table>

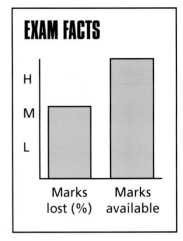
KEY FACTS

- To solve inequalities graphically, draw regions bounded by straight lines.
- A solid line as a boundary is considered to be part of the shaded region, a dashed boundary is not part of the shaded region.

Getting it right

The graph of the straight line with equation $3y + x = 6$ has been drawn on the grid.

$3y + x < 6 \qquad y > 0 \qquad x > 0$

x and y are integers.

On the grid, mark with a cross (×), each of the **two** points which satisfy **all** 3 inequalities.

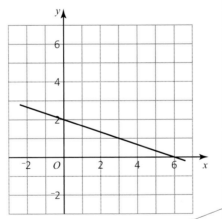

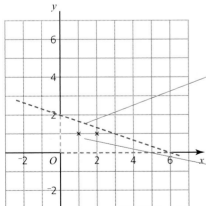

WARNING ⚠️

A common error is to include the points on the boundaries.

Since the inequality signs are either < or > all the boundary lines are dashed.

The coordinates of all points below the red line satisfy the inequality $3y + x < 6$

The coordinates of all points to the right of the green line satisfy the inequality $x > 0$

The coordinates of all points above the blue line satisfy the inequality $y > 0$

The two points in the region with integer coordinates are each marked with a cross.

Show, by shading on the grid, the region which satisfies all three of these inequalities.

$y \leqslant 5$ $y \leqslant 2x$ $y \geqslant x + 1$

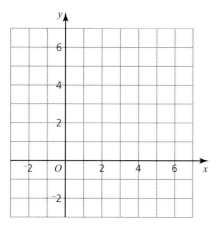

Label your region **R**.

(4400 May 2007)

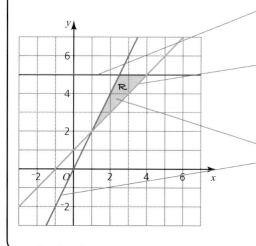

Since the inequality signs are either ⩽ or ⩾ all the boundary lines are solid.

Draw the line $y = 5$. This gets 1 mark. The coordinates of all points on and below the red line satisfy the inequality $y \leqslant 5$

Draw the line $y = x + 1$. This gets 1 mark. The coordinates of all points on and above the green line satisfy the inequality $y \geqslant x + 1$

Draw the line $y = 2x$. This gets 1 mark. The coordinates of all points on and below the blue line satisfy the inequality $y \leqslant 2x$

Shade the region which is below the red line **and** is above the green line **and** is below the blue line. Label this region as **R**. This gets 1 mark

1 The graphs of the straight lines with equations $3y + 2x = 12$ and $y = x - 1$ have been drawn on the grid.

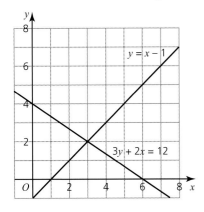

$3y + 2x > 12 \qquad y < x - 1 \qquad x < 6$

x and y are integers.

On the grid, mark with a cross (×), each of the **four** points which satisfy **all** 3 inequalities.

(1388 June 2006)

2 A
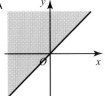
B
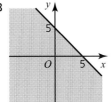

C

D
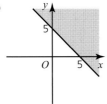

Write down the letter of the graph in which the shaded region represents the inequality

i $y \leqslant x$ ii $x + y \leqslant 5$

(1388 March 2007)

3 On the grid, show, by shading, the region which satisfies all three of the inequalities.

$x < 3$ $y > -2$ $y < x$

Label the region **R**.

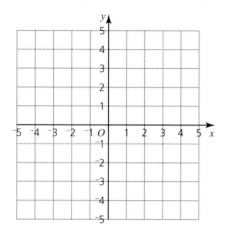

(1387 November 2006)

4

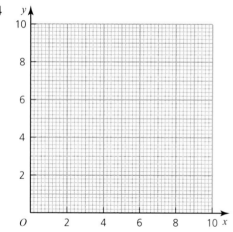

On the grid, shade the region for which

$x + 2y \leqslant 6$ $0 \leqslant x \leqslant 4$ and $y \geqslant 0$

(1384 November 1997)

5

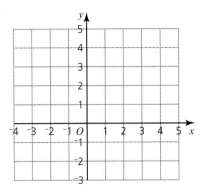

a On the grid, draw the line $x + y = 4$

b On the grid, show clearly the region defined by the inequalities

$x + y \geqslant 4$ $x \leqslant 3$ $y < 4$

(4400 November 2004)

6 Show, by shading on the grid, the region which satisfies these inequalities

$1 \leqslant x \leqslant 3$ **and** $-4 \leqslant y \leqslant -2$

Label your region **R**.

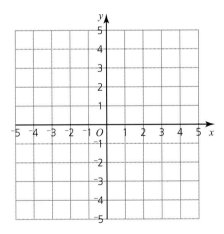

(4400 May 2006)

SKILLS

Solve problems using direct proportion
Solve problems using inverse proportion

EXAM FACTS

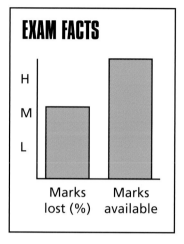

Marks lost (%)   Marks available

KEY FACTS

- If y is directly proportional to x then $y = kx$ where k is a constant
  The notation for 'y is proportional to x' is $y \propto x$
  k is known as the constant of proportionality.

- If y is directly proportional to the square of x then $y = kx^2$ where k is a constant.

- If y is directly proportional to the square root of x then $y = k\sqrt{x}$ where k is a constant.

- If y is inversely proportional to x then $y = \dfrac{k}{x}$ where k is a constant.

- If y is inversely proportional to the square of x then $y = \dfrac{k}{x^2}$ where k is a constant.

- If y is inversely proportional to the square root of x then $y = \dfrac{k}{\sqrt{x}}$ where k is a constant.

Getting it right

Write the relationship as a formula using k.

WARNING

A common error is to leave the answer as $y = kx^2$.

y is directly proportional to the square of x and $y = 2$ when $x = 10$

a Find a formula for y in terms of x.
b Find the value of x when $y = 20$
   Give your answer correct to 3 significant figures.

a $y = kx^2$
   $2 = k \times 10^2$
   $k = \dfrac{1}{50}$
   $y = \dfrac{1}{50}x^2$

b $20 = \dfrac{1}{50} \times x^2$
   $x^2 = 50 \times 20 = 1000$
   $x = \sqrt{1000}$
   $= 31.622...$
   $x = 31.6$ (to 3 s.f.)

Substitute $y = 2$ and $x = 10$ into $y = kx^2$ and work out the value of k.

Replace k by $\dfrac{1}{50}$ in $y = kx^2$ to obtain the formula for y in terms of x.

Substitute $y = 20$ into $y = \dfrac{1}{50}x^2$

p is inversely proportional to q and $p = 6$ when $q = 8$

a Find a formula for p in terms of q.
b Find the value of q when $p = 25$

a $p = \dfrac{k}{q}$

   $6 = \dfrac{k}{8}, \ k = 48$

   $p = \dfrac{48}{q}$

b $25 = \dfrac{48}{q}, \quad q = \dfrac{48}{25} = 1.92$

Write the relationship as a formula using k.

Substitute $p = 6$, $q = 8$ into $p = \dfrac{k}{q}$ and use it to find the value of k.

Replace k by 48 in $p = \dfrac{k}{q}$.

Substitute $p = 25$ into $p = \dfrac{48}{q}$ and find q.

WARNING

A common error is to rearrange $25 = \dfrac{48}{q}$ as $q = \dfrac{25}{48}$

The diameter d cm of a set of cylinders, each with the same volume is inversely proportional to the square root of the height h cm.

When $h = 6.25$, $d = 8$

a Find a formula for d in terms of h.
b $h = 16$, find the value of d.
c $d = 12.5$, find the value of h.

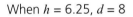

a $d = \dfrac{k}{\sqrt{h}}$

   $8 = \dfrac{k}{\sqrt{6.25}} = \dfrac{k}{2.5}$

   $k = 2.5 \times 8 = 20$

   $d = \dfrac{20}{\sqrt{h}}$

b $d = \dfrac{20}{\sqrt{16}} = \dfrac{20}{4} = 5$

c $12.5 = \dfrac{20}{\sqrt{h}}$

   $12.5 \times \sqrt{h} = 20$

   $\sqrt{h} = \dfrac{20}{12.5} = 1.6$

   $h = 1.6^2 = 2.56$

Write the relationship between d and h as a formula using the constant of proportionality k.
This will get 1 mark.

Substitute $h = 6.25$ and $d = 8$ into $d = \dfrac{k}{\sqrt{h}}$ and find the value of k.
This will get 1 mark.

Replace k by 20 in $d = \dfrac{k}{\sqrt{h}}$.

Substitute $h = 16$ into $d = \dfrac{20}{\sqrt{h}}$.

Substitute $d = 12.5$ into $d = \dfrac{20}{\sqrt{h}}$ and rearrange to the form $\sqrt{h} = \ldots$

WARNING

A common error is to write $h = \sqrt{1.6}$

1 y is directly proportional to x and $y = 10$ when $x = 8$
   a Find a formula for y in terms of x.
   b Find the value of y when $x = 13$

2 y is directly proportional to the square of x and $y = 20$ when $x = 5$
   a Find a formula for y in terms of x.
   b Find the value of y when $x = 6$
   c Find the value of x when $y = 9.8$

3 y is directly proportional to the square root of x and $y = 40$ when $x = 100$
   a Find a formula for y in terms of x.
   b Find the value of x when $y = 16$

4 y is inversely proportional to x and $y = 8$ when $x = 12$
   a Find a formula for y in terms of x.
   b Find the value of y when $x = 2$
   c Find the value of x when $y = 2.5$

5 y is inversely proportional to the square of x and $y = 10$ when $x = 2$
   a Find a formula for y in terms of x.
   b Find the value of x when $y = 20$
      Give your answer correct to 3 significant figures.

6 y is inversely proportional to the square root of x and $y = 12$ when $x = 25$
   a Find a formula for y in terms of x.
   b Find the value of y when $x = 9$
   c Find the value of x when $y = 10$

7 The distance, D metres, travelled by a particle is directly proportional to the square of the time, t seconds, taken.
   When $t = 40$, $D = 30$
   a Find a formula for D in terms of t.
   b Calculate the value of D when $t = 64$
   c Calculate the value of t when $D = 12$
      Give your answer correct to 3 significant figures.
(1387 November 2005)

8 The time, T seconds, it takes a water heater to boil some water is directly proportional to the mass of water, m kg, in the water heater.
   When $m = 250$, $T = 600$

   a Find T when $m = 400$

The time, T seconds, it takes a water heater to boil a constant mass of water is inversely proportional to the power, P watts, of the water heater.
When $P = 1400$, $T = 360$

   b Find the value of T when $P = 900$
(1387 June 2006)

9 A ball falls vertically after being dropped.
The ball falls a distance d metres in a time of t seconds.
d is directly proportional to the square of t.

The ball falls 20 metres in a time of 2 seconds.

   a   Find a formula for d in terms of t.
   b   Calculate the distance the ball falls in 3 seconds.
   c   Calculate the time taken for the ball to fall 605 m.

(1387 November 2006)

10 The number of beats in a minute, n, of a pendulum of length l m is
inversely proportional to the square root of l.
When $l = 0.25$, $n = 60$
   a   Find a formula for n in terms of l.
   b   Calculate the value of n when $l = 0.36$
   c   Calculate the length of a pendulum which makes 20 beats in a minute.

SKILLS

Factorise using the difference of two squares
Factorise expressions of the form $ax^2 + bx + c$, where a, b and c are numbers and $a \neq 1$

EXAM FACTS

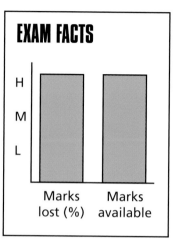

KEY FACTS

- To factorise the difference of two squares use the general rule
  $A^2 - B^2 = (A + B)(A - B)$

- To factorise expressions of the form $ax^2 + bx + c$, for example to factorise $2x^2 + 7x + 6$

  - multiply the coefficient of x^2 (2) by the constant term (+6) which gives +12

  - find two numbers whose product is +12 and whose sum is the coefficient of x (+7). The two numbers are +3 and +4

  - split the x term using these two numbers then factorise by grouping.

Getting it right

Factorise $2x^2 + 7x + 6$

$2x^2 + 7x + 6$
$= 2x^2 + 4x + 3x + 6$
$= 2x(x + 2) + 3(x + 2)$
$= (x + 2)(2x + 3)$

> Multiply the coefficient of x^2 (2) by the constant term (+6) gives +12. Find two numbers whose product is +12 and whose sum is +7

> Factorise $2x^2 + 4x + 3x + 6$ by grouping.

> Check your answer by expanding $(x + 2)(2x + 3)$.

Factorise completely $3mp^2 - 12m^3$

$3mp^2 - 12m^3$
$= 3m(p^2 - 4m^2)$
$= 3m[(p)^2 - (2m)^2]$
$= 3m(p + 2m)(p - 2m)$

> Take out the common factor $3m$.
> The result $3m(p^2 - 4m^2)$ is not yet completely factorised.

> Use $A^2 - B^2 = (A + B)(A - B)$ with $A = p$ and $B = 2m$

a Factorise $p^2 - q^2$

b Hence work out the value of $101^2 - 99^2$

a $p^2 - q^2 = (p + q)(p - q)$

Use $A^2 - B^2 = (A + B)(A - B)$

b $101^2 - 99^2 = (101 + 99)(101 - 99)$
$$= 200 \times 2$$
$$= 400$$

Use $A^2 - B^2 = (A + B)(A - B)$
with $A = 101$ and $B = 99$

Now try these

1 Factorise
a $y^2 - 9$
b $121t^2 - 1$
c $p^2 - 16q^2$
d $25 - a^2$

2 Factorise
a $2x^2 + 7x + 3$
b $5x^2 - 8x + 3$
c $5x^2 - 3x - 2$
d $3x^2 - 2x - 8$

3 Factorise completely $2x^2 - 50y^2$

4 a Factorise $m^2 - n^2$
b Hence work out the value of $550^2 - 450^2$

5 a Factorise $6x^2 + 7x - 3$
b Factorise $9x^2 - 6x + 1$
c Write down a common factor of $6x^2 + 7x - 3$ and $9x^2 - 6x + 1$

6 Find the common factor of $4x^2 - 9$ and $4x^2 + 2x - 12$

7 Show that $(2a - 1)^2 - (2b - 1)^2 = 4(a - b)(a + b - 1)$

(1387 June 2003)

8 a Factorise $2x^2 - 7x + 6$
b i Factorise fully $(n^2 - a^2) - (n - a)^2$
 n and a are integers.
 ii Explain why $(n^2 - a^2) - (n - a)^2$ is always an even integer.

(1387 June 2006)

SKILL

Solve quadratic equations using the formula

EXAM FACTS

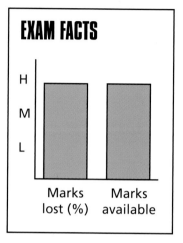

Marks lost (%)   Marks available

KEY FACTS

- A quadratic equation is one which can be written as $ax^2 + bx + c = 0$ where a, b and c are numbers and a is not zero.
- The solutions of the quadratic equation $ax^2 + bx + c = 0$ are given by

$$x = \frac{-b \pm \sqrt{b^2 - 4ac}}{2a}$$

This formula is given on the formula sheet on the exam paper.

Getting it right

Solve the equation $x^3 + 5x + 3 = 0$
Give your answers correct to 3 significant figures.

$a = 1, b = 5, c = 3$

$$x = \frac{-5 \pm \sqrt{5^2 - 4 \times 1 \times 3}}{2 \times 1}$$

$$x = \frac{-5 \pm \sqrt{25 - 12}}{2}$$

$$x = \frac{-5 \pm \sqrt{13}}{2}$$

$$x = \frac{-5 \pm 3.60555...}{2}$$

(+ sign)

$$x = \frac{-5 + 3.60555...}{2}$$

$$x = \frac{-1.39444...}{2}$$

$x = -0.697$ (to 3 s.f.)

> Asking for 3 s.f. is a hint that the left-hand side of the equation will not factorise.

> Comparing $x^2 + 5x + 3 = 0$ with $ax^2 + bx + c = 0$

> Substitute in the formula for a, b and c. This would get 1 mark.

> The $\pm$ sign means that one solution of the equation comes from using the + sign, the other from the − sign.

> Work out each term separately, until the answer is of this form. This would get 1 mark.

> Write down the square root correct to at least 4 significant figures. Use this value to work out the solutions correct to 3 significant figures.

> Find one solution using the + sign of the $\pm$

> Work out the numerator of the expression first and write it down. Then write the value of x correct to 3 significant figures.

Repeat for the other solution using the − sign.

(− sign)

$$x = \frac{-5 - 3.60555...}{2}$$

$$x = \frac{-8.60555...}{2}$$

$$x = -4.30 \text{ (to 3 s.f.)}$$

EXAM TIP

In general, all working should be to at least one more significant figure than the degree of accuracy specified in the question.

Solve the equation $2x^2 - 3x = 1$
Give your answers correct to 3 significant figures.

$$2x^2 - 3x - 1 = 0$$

$$a = 2, b = (-3), c = (-1)$$

$$x = \frac{-(-3) \pm \sqrt{(-3)^2 - 4 \times 2 \times (-1)}}{2 \times 2}$$

$$x = \frac{3 \pm \sqrt{9 + 8}}{4}$$

$$x = \frac{3 \pm \sqrt{17}}{4}$$

$$x = \frac{3 + 4.12310...}{4} = 1.78 \text{ (to 3 s.f.)}$$

$$x = \frac{3 - 4.12310...}{4} = -0.281 \text{ (to 3 s.f.)}$$

Rearrange the equation to the form $ax^2 + bx + c = 0$

List the values of a, b and c. Put any negative values in brackets.

Substitute for a, b and c in the formula. Make sure that any negative values are still in brackets.

WARNING

Errors with signs are common.

Write down the square root correct to at least 4 significant figures. Use this value to work out the solutions correct to 3 significant figures.

Now try these

Solve each of the following quadratic equations
In each case, give your solutions correct to 3 significant figures.

1 $x^2 + 4x + 2 = 0$

2 $x^2 + 5x + 3 = 0$

3 $2x^2 + 4x + 1 = 0$

4 $x^2 - 5x - 8 = 0$

(1387 June 2006)

5 $3x^2 - 8x + 2 = 0$

6 $x^2 - 7x - 3 = 0$

7 $x^2 + 2x - 1 = 0$

8 $x^2 + 6x = 3$

9 $x^2 - 2 = 4x$

10 $4x(x - 1) = 1$

11 $(x + 1)(x - 2) = 3$

12 $3x + 2 = 2x^2$

13 $5(x + 4) = x(2x + 1)$

14 $6x^2 - 5 = 3x^2 + 5$

Solving problems using quadratic equations

SKILLS

Solve quadratic equations using a variety of methods

Interpret solutions of quadratic equations in the context of a problem

KEY FACTS

- A quadratic equation generally has 2 solutions.

- Some problems lead to quadratic equations.
  It is possible that one of the solutions to the equation is not a valid solution to the original problem.

EXAM FACTS

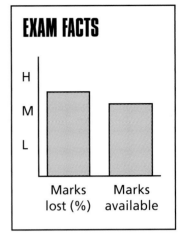

REFERENCE

For a reminder of how to solve quadratic equations using the formula, turn to pages 44 and 45.

Getting it right

$2x + 1$

$x + 4$

Diagram **NOT** accurately drawn

The diagram shows a rectangle. All the measurements are in centimetres.
x satisfies the equation $2x^2 + 9x - 68 = 0$

a Solve the equation.
b Find the area of the rectangle.

a $2x^2 + 9x - 68 = 0$

   $(2x + 17)(x - 4) = 0$

   $x = -\dfrac{17}{2}$ or $x = 4$

b Area $= (2x + 1)(x + 4)$

   when $x = 4$, Area $= 9 \times 8 = 72 \text{ cm}^2$

If a degree of accuracy is not specified in the question, the quadratic equation can be often solved by factorisation.

$2x^2$ means that the factors will start $2x$ and x
Pairs of factors of 68 are 1 and 68, 2 and 34 and 4 and 17

The equation has two solutions. Both solutions must be written down even if one of them cannot be a solution of the problem.

Lengths cannot be negative and so $x = 4$ is the only valid answer.

The diagram shows a 6-sided shape.
All the corners are right angles.
All measurements are given in centimetres.

The area of the shape is 25 cm².
It can be shown that $6x^2 + 17x - 39 = 0$

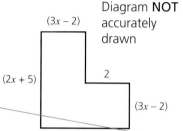

Diagram **NOT** accurately drawn

a Solve the equation $6x^2 + 17x - 39 = 0$
b Find the length of the longest side of the shape.

REFERENCE

This equation is derived on page 59.

a $6x^2 + 17x - 39 = 0$
   $(3x + 13)(2x - 3) = 0$
   $x = -\dfrac{13}{3}$ or $x = \dfrac{3}{2}$

$6 \times 39 = 2 \times 3 \times 3 \times 13$
$ = 2 \times 13 \times 3 \times 3$
$ = 26 \times 9$
$26 - 9 = 17$

b When $x = \dfrac{3}{2}$, $2x + 5 = 8$ and $3x = 4.5$
   The length of the longest side is 8 cm.

Lengths cannot be negative.

From the diagram, it is clear that the longest side is either the height $2x + 5$ or the base $3x$.

The distance from A to B is 24 kilometres.
Kavish cycles from A to B at a speed of x kilometres per hour.
He cycles back from B to A at a speed of $(x + 4)$ kilometres per hour.
The total time for the journey from A to B and back is 5 hours.
x satisfies the quadratic equation $5x^2 - 28x - 96 = 0$

a Solve the equation.
b Find the speed that Kavish cycled back from B to A.

a $5x^2 - 28x - 96 = 0$
   $(5x + 12)(x - 8) = 0$
   $x = -\dfrac{12}{5}$ or $x = 8$

$5 \times 96 = 5 \times 12 \times 8$
$ = 5 \times 8 \times 12$
$ = 40 \times 12$
$40 - 12 = 28$

Speed cannot be negative.

b Speed > 0 so $x = 8$
   Speed that Kavish cycled back from B to $A = x + 4$
   $= 12$ kilometres per hour.

1 The product of the positive numbers x and $x + 4$ is 45
  It can be shown that $x^2 + 4x - 45 = 0$

  a   Solve the equation $x^2 + 4x - 45 = 0$
  b   Write down the larger of the two positive numbers.

2
  Diagram **NOT** accurately drawn

  In the diagram, all the measurements are in centimetres.
  It can be shown that $x^2 + 2x - 15 = 0$

  a   Solve this equation.
  b   Find the area of the triangle.

3
  Diagram **NOT** accurately drawn

  In the rectangle all the measurements are in centimetres.
  The area of the rectangle is 15 cm^2.
  It can be shown that $x^2 + 14x + 48 = 15$

  a   Solve the equation $x^2 + 14x + 48 = 15$
  b   Find the length of each side.

4 Jim walks a distance of 18 miles at a speed of x mph. Bill walks the same
  distance at a speed of $(x + 2)$ mph. Bill takes $1\frac{1}{2}$ hours less than Jim.
  It can be shown that $x^2 + 2x - 24 = 0$

  a   Solve $x^2 + 2x - 24 = 0$
  b   Find the time that Bill took.

5 The dimensions of a cuboid are 2 cm, $(x + 4)$ cm and $(6 - x)$ cm.
  The volume of the cuboid is 32 cm^3.
  It can be shown that $x^2 - 2x - 8 = 0$

  a   Solve $x^2 - 2x - 8 = 0$
  b   Use both solutions of the equation to find the dimensions of the cuboid.

6 Beccy has £90 to spend on CDs. The normal price of a CD is £x
  In a sale, all prices of CDs are reduced by £1
  If she spends £90 on CDs at the sale price, she can buy 1 more CD than if
  she spends £90 at the normal price.
  It can be shown that $x^2 - x = 90$

  a   Solve the equation $x^2 - x = 90$
  b   Work out how many CDs Beccy can buy at the sale price.

7

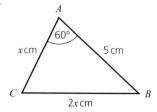

Diagram **NOT** accurately drawn

In triangle ABC, $AB = 5$ cm, $AC = x$ cm, $BC = 2x$ cm and angle $BAC = 60°$.
It can be shown that $3x^2 + 5x - 25 = 0$

a   Solve the equation $3x^2 + 5x - 25 = 0$
    Give your answers correct to 4 significant figures.

b   Find the length of BC. Give your answer correct to 3 significant figures.

8 Peter cuts a square out of a rectangular piece of metal.

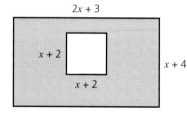

Diagram **NOT** accurately drawn

The length of the rectangle is $2x + 3$
The width of the rectangle is $x + 4$
The length of the side of the square is $x + 2$
All measurements are in centimetres.

The shaded shape in the diagram shows the metal remaining.
The area of the shaded shape is 20 cm^2.
It can be shown that $x^2 + 7x - 12 = 0$

a   Solve the equation $x^2 + 7x - 12 = 0$
    Give your answers correct to 4 significant figures.

b   Hence, find the perimeter of the square.
    Give your answer correct to 3 significant figures.

13 Completing the square

SKILLS

Complete the square for a quadratic expression
Use quadratic expressions in completed square form to solve equations
Find the minimum value of a quadratic function
Find the maximum value of a quadratic function

EXAM FACTS

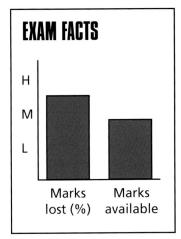

KEY FACTS

- Any quadratic function of the form $x^2 + bx + c$ can be written as $(x + p)^2 + q$ where b, c, p and q are numbers.

- Any quadratic function of the form $c + bx - x^2$ can be written as $q - (x + p)^2$ where b, c, p and q are numbers.

- The minimum value of $(x + p)^2 + q$ is q and occurs when $x = -p$ (Because $(x + p)^2$ is a square its value is never negative and its minimum value is 0, occurring when $x = -p$).

- The maximum value of $q - (x + p)^2$ is q and occurs when $x = -p$.

Getting it right

Write $x^2 + 8x + 22$ in the form $(x + p)^2 + q$ where p and q are numbers.

$(x + p)^2 + q = x^2 + 2px + p^2 + q$

Comparing $x^2 + 2px + p^2 + q$ with $x^2 + 8x + 22$

$2p = 8$ and $p^2 + q = 22$
$p = 4$ and $4^2 + q = 22$
$q = 6$

so $x^2 + 8x + 22 = (x + p)^2 + q$
$= (x + 4)^2 + 6$

WARNING

A common error is to write $(x + p)^2 = x^2 + p^2$.

The terms in x have to be the same. The constants have to be the same.

The value of p is always half the coefficient of x.

a Write $x^2 - 6x + 1$ in the form $(x - p)^2 + q$.
b Use your answer to part **a** to solve $x^2 - 6x + 1 = 0$
Give your answer in the form $a \pm \sqrt{b}$ where a and b are integers.

a $x^2 - 6x + 1 = x^2 - 2px + p^2 + q$
coeff of x: $-6 = -2p$, so $p = 3$
constant: $1 = 3^2 + q$, so $q = -8$
so $x^2 - 6x + 1 = (x - 3)^2 - 8$

b The equation can be written as
$(x - 3)^2 - 8 = 0$
$(x - 3)^2 = 8$
$x - 3 = \pm \sqrt{8}$
$x = 3 \pm \sqrt{8}$

Finding the value of p and the value of q is sufficient to get the marks.

The expression $8x - x^2$ can be written in the form $p - (x - q)^2$, for all values of x.
a Find the value of p and the value of q.

The expression $8x - x^2$ has a maximum value.
b State the value of x for which this maximum value occurs.

a $p - (x - q)^2 = p - (x^2 - 2qx + q^2)$
$= p - x^2 + 2qx - q^2$
$= p - q^2 + 2qx - x^2$

Comparing $p - q^2 + 2qx - x^2$ with $8x - x^2$
$2q = 8$ and $p - q^2 = 0$
$q = 4$ and $p = 16$

b $8x - x^2 = 16 - (x - 4)^2$

The maximum value of $8x - x^2$ is 16
and occurs when $x = 4$

This means that numerical values of p and q have to be found so that the expansion and simplification of $p - (x - q)^2$ would give $8x - x^2$.

WARNING

A common error is to forget to use brackets and to write $p - x^2 - 2qx + q^2$.

Now try these

1 Write $x^2 + 8x + 17$ in the form $(x + p)^2 + q$.

2 Write $x^2 + 4x$ in the form $(x + p)^2 + q$.

3 Write $x^2 - 8x + 1 = 0$ in the form $(x + p)^2 + q$.

4 Write $x^2 - x + 1$ in the form $(x + p)^2 + q$.

5 The expression $x^2 - 6x + 14$ can be written in the form $(x - p)^2 + q$, for all values of x.
  a Find the value of i p, ii q.
     The equation of a curve is $y = f(x)$, where $f(x) = x^2 - 6x + 14$

Here is a sketch of the graph of $y = f(x)$.

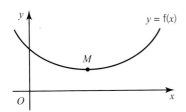

  b Write down the coordinates of the minimum point, M, of the curve.
(1387 November 2003)

6 Show that $x^2 - 4x + 15$ can be written as $(x + p)^2 + q$ for all values of x.
State the values of p and q.
(1387 November 2005)

7 a Write $x^2 + 6x$ in the form $(x + p)^2 + q$, stating the values of p and q.
  b Use your answer to part **a** to solve the equation $x^2 + 6x = 4$
     Give your answers in the form $p \pm \sqrt{q}$, where p and q are integers.

8 Given that $x^2 - 14x + a = (x + b)^2$ for all values of x, find the value of a and the value of b.
(1388 November 2005)

9 Write $x^2 - 6x + 3$ in the form $(x + p)^2 + q$.
Hence, find the minimum value of $x^2 - 6x + 3$

10 Show that $20 - 10x - x^2$ can be written in the form $q - (p - x)^2$.
Hence, find the maximum value of $20 - 10x - x^2$ and the value of x at which this maximum occurs.

Simplifying algebraic fractions

SKILLS

Simplifying rational expressions
Adding and subtracting rational expressions

EXAM FACTS

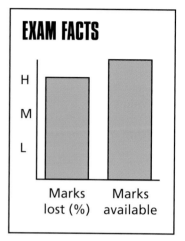

KEY FACTS

- To simplify rational expressions, factorise both the numerator and denominator and cancel any common factors.
- To add or subtract algebraic fractions with different denominators, apply these steps
    - factorise the denominators if possible,
    - write each fraction as a fraction with a common denominator,
    - add or subtract the fractions and factorise the numerator if possible,
    - simplify the algebraic fraction if possible.

REFERENCE

For a reminder of how to factorise turn to pages 42 and 43.

Getting it right

Simplify fully $\dfrac{x^2 + 7x + 10}{x^2 - 25}$

(1388 March 2007)

$$\frac{x^2 + 7x + 10}{x^2 - 25} = \frac{(x + 2)(x + 5)}{(x - 5)(x + 5)}$$

$$= \frac{(x + 2)(x + 5)}{(x - 5)(x + 5)}$$

$$= \frac{(x + 2)}{(x - 5)}$$

WARNING

A common error is to cancel part of the expression in the numerator by part of the expression in the denominator, for example, cancelling the x^2 terms.

Correct factorisation of $x^2 + 7x + 10$ gets 1 mark.

Correct factorisation of $x^2 - 25$ gets 1 mark.

1 mark for the correct answer.

Write as a single fraction $\dfrac{1}{2x + 8} + \dfrac{2}{x^2 + 4x}$

$$\dfrac{1}{2x + 8} + \dfrac{2}{x^2 + 4x} = \dfrac{1}{2(x + 4)} + \dfrac{2}{x(x + 4)}$$

$$= \dfrac{x}{2x(x + 4)} + \dfrac{4}{2x(x + 4)}$$

$$= \dfrac{x + 4}{2x(x + 4)}$$

$$= \dfrac{(x + 4)}{2x(x + 4)}$$

$$= \dfrac{1}{2x}$$

Factorising the denominators gets 1 mark.

Writing each fraction as a fraction with a common denominator gets 1 mark.

Adding the fractions gets 1 mark.

Simplifying the fraction to give the correct answer gets 1 mark.

Now try these

1 Simplify $\dfrac{(x - 1)^2}{x - 1}$ *(1388 June 2005)*

2 Write as a single fraction $\dfrac{2}{x} + \dfrac{3}{2x}$

3 Simplify fully $\dfrac{3x + 6}{x^2 - 4}$ *(2544 March 2007)*

4 Simplify fully $\dfrac{36 - x^2}{36 + 6x}$

5 Write as a single fraction $\dfrac{4}{x(x + 3)} + \dfrac{5}{(x + 3)}$ *(2544 March 2007)*

6 Simplify fully $\dfrac{x^2 + 5x + 6}{x^2 + 2x}$ *(1387 June 2006)*

7 Simplify fully $\dfrac{y^2 - 4y}{y^2 - 6y + 8}$

8 Simplify fully $\dfrac{x^2 - 9}{x^2 - 9x + 18}$ *(4400 November 2004)*

9 Express the algebraic fraction $\dfrac{2x^2 - 3x - 20}{x^2 - 16}$ as simply as possible.
                                            (4400 May 2004)

10 Express as a single fraction $\dfrac{1}{2n - 1} + \dfrac{1}{2n + 1}$ *(1385 November 1999)*

11 Simplify $\dfrac{1}{(x + 2)} - \dfrac{3}{(x - 1)}$ *(1384 November 1997)*

12 Simplify fully $\dfrac{2}{x - 1} + \dfrac{x - 11}{x^2 + 3x - 4}$ *(4400 May 2005)*

SKILLS

Use graphs to find approximate solutions of equations

Use algebra to find exact solutions of equations

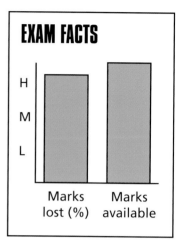

Marks lost (%)   Marks available

KEY FACTS

- The equation of a straight line with gradient m and intercept c is $y = mx + c$.

- The equation of a straight line can also be written as $ax + by = c$ where a, b and c are numbers.
  (These types of equation are known as linear equations.)

- The equation of a circle with radius r and centre O is $x^2 + y^2 = r^2$.

- The exact coordinates of the point(s) of intersection of the straight line $y = mx + c$ and the equation of any curve are found by solving the equation of the line and the equation of the curve simultaneously.

- Estimates of the solutions of the simultaneous equations $y = mx + c$ (or $ax + by = c$) and $x^2 + y^2 = r^2$ can be found graphically from the coordinates of the point(s) of intersection of the straight line and the circle.

Getting it right

Solve the simultaneous equations $y = 4x - 3$ and $y = 3x^2 - 2$

Substitute for y from the linear equation.

$$4x - 3 = 3x^2 - 2$$
$$3x^2 - 4x + 1 = 0$$
$$(3x - 1)(x - 1) = 0$$

$x = \frac{1}{3}$ or $x = 1$

When $x = \frac{1}{3}$, $y = 4 \times \frac{1}{3} - 3 = -\frac{5}{3}$
When $x = 1$, $y = 4 \times 1 - 3 = 1$

WARNING

A common error is to fail to pair each y value with the corresponding x value.

Substitute into the linear equation as this is easier.

a On the grid, draw the graph of $x^2 + y^2 = 16$
b By drawing a suitable straight line on the grid, find estimates of the solutions of the simultaneous equations
$x^2 + y^2 = 16$
$x + 2y = 6$

a

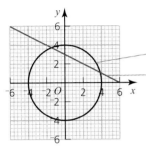

Use compasses to draw the circle $x^2 + y^2 = 16$
The radius is $\sqrt{16}$

Draw the straight line with equation $x + 2y = 6$ on the grid by plotting two points that lie on the line and then using a ruler to draw the straight line through them.

b The solution of the equations can be found from the points of intersection of the two graphs.

The graphs intersect at $(-1.4, 3.8)$ and $(3.8, 1.1)$

So the solutions of the equations are $x = -1.4$, $y = 3.8$ and $x = 3.8$, $y = 1.1$

EXAM TIP

When the answers are read from a graph answers will be accepted within a given range e.g. $-1.5 \leqslant x \leqslant -1.3$

Find the exact coordinates of the points of intersection of the line $y = 2x - 1$ and the circle $x^2 + y^2 = 34$

Substitute for y from the equation of the straight line into the equation of the circle.

$x^2 + (2x - 1)^2 = 34$
$x^2 + 4x^2 - 4x + 1 = 34$
$5x^2 - 4x - 33 = 0$
$(5x + 11)(x - 3) = 0$
$x = -2.2$ or $x = 3$

When $x = -2.2$, $y = 2 \times (-2.2) - 1 = -5.4$
When $x = 3$, $y = 2 \times 3 - 1 = 5$

The coordinates of the points of intersection are $(-2.2, -5.4)$ and $(3, 5)$

y is already the subject.

WARNING

A common error is to omit the $-4x$ term in the expansion of $(2x - 1)^2$

Use the linear equation to find the value of y for each value of x.

Now try these

1 Solve $y = 3x$, $y = 2x^2 + 1$

2 Solve $y = 4$, $y = x^2 - 3x$

3 Find algebraically the points of intersection of the straight line $y = x$ and the curve $y = x^2 - 6$

4 The line $y = 3x + 1$ intersects the curve $y = 4x^2$ at the points A and B. Find the coordinates of A and B.

5 a   Draw the circle with equation $x^2 + y^2 = 25$
   b   Draw the straight line with equation $x + y = 3$
      Hence, find estimates of the solutions of the equations $x^2 + y^2 = 25$
$$x + y = 3$$

6 Solve $x^2 + y^2 = 5$, $y = x - 3$

7 Solve $x^2 + y^2 = 26$, $y = x - 4$

8 The line $x = 2$ intersects the circle $x^2 + y^2 = 13$ at the points P and Q. Find the exact length of PQ.

9 By eliminating y, find the solutions to the simultaneous equations
$y - 2x = 3$
$x^2 + y^2 = 18$ *(1388 June 2006)*

10 The diagram shows a circle of radius 5, centre the origin.

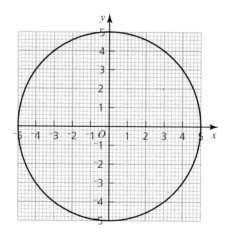

Draw a suitable straight line on the diagram to find estimates of the solutions to the pair of equations

$x^2 + y^2 = 25$
$y = 2x + 1$

(1387 November 2005)

16 Algebraic proofs

EXAM FACTS

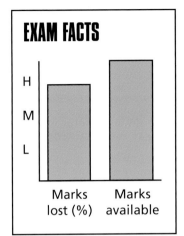

Getting it right

The diagram below shows a 6-sided shape.
All the corners are right angles.
All measurements are given in centimetres.

The area of the shape is 25 cm².
Show that $6x^2 + 17x - 39 = 0$

Diagram **NOT** accurately drawn

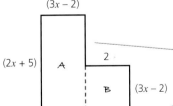

(1387 June 2005)

> **WARNING** ⚠️
>
> A common error is to attempt to solve the given equation and then use the solutions to find the lengths in the diagram. Verifying that the area of the shape is then 25 cm² is not a proof and gets no marks.

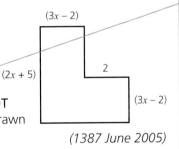

Area $A = (2x + 5)(3x - 2)$
Area $B = 2(3x - 2)$

> Split the shape into two rectangles.

> This expression gets 1 mark.

$$\text{Total area} = (2x + 5)(3x - 2) + 2(3x - 2)$$
$$= 6x^2 - 4x + 15x - 10 + 6x - 4$$
$$= 6x^2 + 17x - 14$$

> Correct expansions get 1 mark.

Area $= 25$, so $\quad 6x^2 + 17x - 14 = 25$
so, $\qquad\qquad\quad 6x^2 + 17x - 39 = 0$

> 1 mark for completing the proof.

Prove algebraically that the sum of the squares of any two consecutive even integers is never a multiple of 8.

For any integer n,
Even number = $2n$
Next even number = $2n + 2$

$(2n)^2 + (2n + 2)^2 = 2n \times 2n + (2n + 2)(2n + 2)$
$= 4n^2 + 4n^2 + 4n + 4n + 4$
$= 8n^2 + 8n + 4$
$= 8(n^2 + n) + 4$
$= (\text{multiple of 8}) + 4$

which is not a multiple of 8

Writing algebraic expressions for two consecutive even numbers gets 1 mark.

$(2n)^2 + (2n + 2)^2$ gets 1 mark.

This gets 1 mark.

1 mark for completing the proof.

Sophie says, 'For any whole number, n, the value of $6n - 1$ is always a prime number.'
Sophie is wrong.
Give an example to show that Sophie is wrong.

For $n = 6$, $6n - 1 = 36 - 1$
$6n - 1 = 35$

35 is not a prime number so
Sophie is wrong.

$n = 6$ gets 1 mark.

Finding any counter example is sufficient to disprove Sophie's statement. There are other possible values of n, for example $n = 11$

This gets 1 mark.

Now try these

1 The diagram shows a prism.
The cross-section of the prism is a right-angled triangle.
The lengths of the sides of the triangle are $3x$ cm, $4x$ cm and $5x$ cm.
The total length of all the edges of the prism is E cm.

HINT

Write a formula for E in terms of x and L.

Show that the length, L cm, of the prism is given by the formula
$L = \frac{1}{3}(E - 24x)$.

(1386 June 2000)

2 The diagram shows a rectangle.

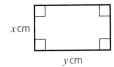

Diagram **NOT**
accurately drawn

The width of the rectangle is x cm and its length is y cm.
The perimeter of the rectangle is 10 cm.

a Show that $x + y = 5$

The length of the diagonal of the rectangle is 4 cm.

b Show that $2x^2 - 10x + 9 = 0$

(1387 November 2006)

3 Sam says, 'For any positive integer, n, the value of $2n + 1$ is never a square number.' Sam is wrong. Give an example to show that Sam is wrong.

4 'The sum of any two prime numbers is never a prime number.'
Show, by means of a counter example, that this statement is not true.

5 By writing the nth term of the sequence 1, 3, 5, 7, ... as $(2n - 1)$,
or otherwise, show that the difference between the squares of any two consecutive odd numbers is a multiple of 8

(1386 June 1999)

6 Prove that $(n + 1)^2 - (n - 1)^2$ is a multiple of 4 for all positive integer values of n.

(1387 November 2003)

7 In triangle ABC, $AB = 5$ cm, $AC = x$ cm,
$BC = 2x$ cm and angle $BAC = 60°$.

Show that $3x^2 + 5x - 25 = 0$

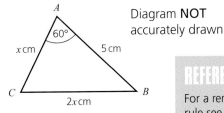

Diagram **NOT**
accurately drawn

REFERENCE

For a reminder of the cosine rule see pages 89 and 90.

HINT

Use the cosine rule.

(1384 November 1997)

8 a Factorise $p^2 - q^2$.

Here is a sequence of numbers

0   3   8   15   24   35   48   .....

b Write down an expression for the nth term of this sequence.

c Show algebraically, that the product of any two consecutive terms of the sequence

0   3   8   15   24   35   48   .....

can be written as the product of four consecutive integers.

(1386 June 2000)

17 Similar triangles

SKILL

Compare corresponding sides in similar triangles to calculate unknown lengths

EXAM FACTS

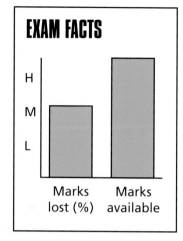

KEY FACTS

- Similar triangles have the same shape but not the same size.
- Similar triangles have equal angles.
- If two triangles are similar, the lengths of pairs of corresponding sides are in the same proportion or ratio.

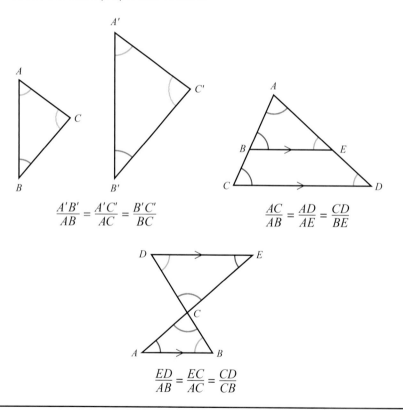

$$\frac{A'B'}{AB} = \frac{A'C'}{AC} = \frac{B'C''}{BC}$$

$$\frac{AC}{AB} = \frac{AD}{AE} = \frac{CD}{BE}$$

$$\frac{ED}{AB} = \frac{EC}{AC} = \frac{CD}{CB}$$

Getting it right

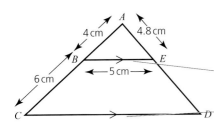

Diagram **NOT** accurately drawn

BE is parallel to CD.
ABC and AED are straight lines.
$AB = 4$ cm, $BC = 6$ cm, $BE = 5$ cm, $AE = 4.8$ cm.

a   Calculate the length of CD.
b   Calculate the length of ED.

(1387 November 2004)

a   Triangles ABE and ACD are similar.

so, $\dfrac{CD}{BE} = \dfrac{AC}{AB}$

$\dfrac{CD}{5} = \dfrac{10}{4}$

$CD = \dfrac{5 \times 10}{4} = 12.5\,cm$

b $\dfrac{AD}{AE} = \dfrac{AC}{AB}$

$\dfrac{AD}{4.8} = \dfrac{10}{4}$

$AD = \dfrac{4.8 \times 10}{4} = 12\,cm$

$ED = 12 - 4.8 = 7.2\,cm$

EXAM TIP

Diagram NOT accurately drawn means that taking measurements from the diagram will give WRONG answers.

Parallel lines show triangle ABE and triangle ACD are similar.

The lengths of corresponding sides are in the same proportion. Put the required side as the numerator of the left-hand side.

Substitute the known lengths. This scores 1 mark.

Rearrange the equation to work out the length of CD.

EXAM TIP

You could also use $\dfrac{AD}{AE} = \dfrac{CD}{BE}$ but it is safer not to in case you have miscalculated the length of CD in part **a**.

This scores 1 mark.

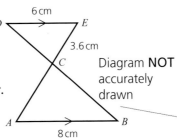

AB is parallel to DE.
ACE and BCD are straight lines.
Calculate the length of AC.

Diagram **NOT** accurately drawn

Triangles CED and CAB are similar.

so, $\dfrac{AC}{EC} = \dfrac{AB}{ED}$

$\dfrac{AC}{3.6} = \dfrac{8}{6}$

$AC = \dfrac{3.6 \times 8}{6} = 4.8\,cm$

Triangle CAB and triangle CED are similar, so the pairs of corresponding sides are
CA and CE
AB and ED
CB and CD
but only the first two pairs are needed in this question.

This scores 1 mark.

1 Triangles *PQR* and *STU* are similar.
Calculate the lengths of
  a *SU*,   **b** *PQ*.

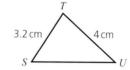

2 The two triangles *ABC* and *PQR* are mathematically similar.
Angle *A* = angle *P*. Angle *B* = angle *Q*.
AB = 8 cm, *AC* = 26 cm, *PQ* = 12 cm, *QR* = 45 cm.

  a   Work out the length of *PR*.
  b   Work out the length of *BC*.

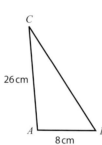
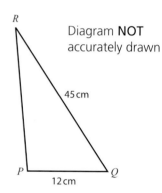

Diagram **NOT**
accurately drawn

(1387 June 2006)

3 *BE* is parallel to *CD*.
  AB = 4.5 cm, *AE* = 5 cm, *ED* = 3 cm, *CD* = 5.6 cm.

  a   Calculate the length of *BE*.
  b   Calculate the length of *BC*.

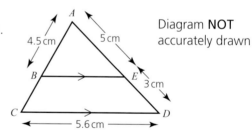

Diagram **NOT**
accurately drawn

(4400 November 2004)

4 In the diagram, *FG* = 5.6 metres, *EH* = 3.5 metres
and *DH* = 15 metres.
EH is parallel to *FG*. *FED* and *DHG* are straight lines.
Calculate the length of *DG*.

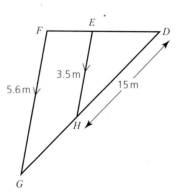

Diagram **NOT**
accurately drawn

(1384 November 1996)

5 *BE* is parallel to *CD*.
$AE = 6$ cm, $ED = 4$ cm, $AB = 4.5$ cm, $BE = 4.8$ cm.

   a Calculate the length of *CD*.

   b Calculate the perimeter of the trapezium *EBCD*.

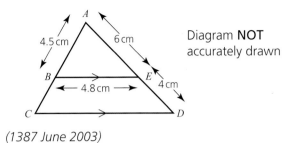

Diagram **NOT** accurately drawn

(1387 June 2003)

6 *AB* is parallel to *DE*. *ACE* and *BCD* are straight lines.
$AB = 6$ cm, $AC = 8$ cm, $CD = 13.5$ cm, $DE = 9$ cm.

   a Work out the length of *CE*.

   b Work out the length of *BC*.

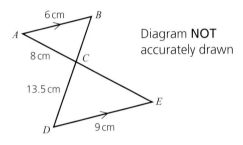

Diagram **NOT** accurately drawn

(1387 November 2005)

7

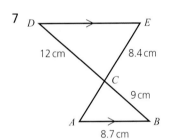

Diagram **NOT** accurately drawn

AB is parallel to *DE*. *ACE* and *BCD* are straight lines.

   a Work out the length of *DE*.

   b Work out the length of *AC*.

8

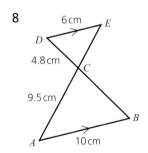

AB is parallel to *DE*. *ACE* and *BCD* are straight lines.

   a Work out the length of *BC*.

   b Work out the length of *CE*.

SKILL

Use ratio of length = $a : b$, ratio of area = $a^2 : b^2$
and ratio of volume = $a^3 : b^3$

EXAM FACTS

Bar chart with axes labelled H, M, L and columns for "Marks lost (%)" and "Marks available"

KEY FACTS

For similar shapes,

- when lengths are multiplied by k, area is multiplied by k^2.

  For example, if the lengths of a shape are multiplied by 4, its area is multiplied by 4^2, that is 16

- when lengths are multiplied by k, volume is multiplied by k^3.

  For example, if the lengths of a shape are multiplied by 4, its volume is multiplied by 4^3, that is 64

Getting it right

Quadrilaterals **P** and **Q** are similar.
The area of quadrilateral **P** is 20 cm^2.
Calculate the area of quadrilateral **Q**.

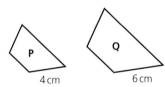

P   4 cm   Q   6 cm

Scale factor = $\frac{6}{4}$ = 1.5

Calculate the number by which lengths have been multiplied, that is, the scale factor.

$1.5^2 = 2.25$

Square the scale factor to find the number by which the area has to be multiplied. This scores 1 mark.

20 cm^2 × 2.25 = 45 cm^2

Multiply the area of quadrilateral **P** by 2.25 to find the area of quadrilateral **Q**. 1 mark for the correct answer.

WARNING ⚠

A common error is to multiply the area by the scale factor, rather than multiplying it by the **square** of the scale factor.

Two cuboids, **S** and **T**, are mathematically similar.

The width of cuboid **S** is 6 cm and the width of cuboid **T** is 30 cm.

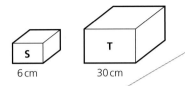

S 6 cm   T 30 cm

a The total surface area of cuboid **S** is 248 cm². Calculate the total surface area of cuboid **T**.
b The volume of cuboid **T** is 30 000 cm³. Calculate the volume of cuboid **S**.

Scale factor = $\frac{30}{6}$ = 5

a 5^2 = 25
   248 cm² × 25 = 6200 cm²

b 5^3 = 125
   30 000 cm³ ÷ 125 = 240 cm³

Calculate the scale factor. There is 1 mark for this.

Square the scale factor to find the number by which the area has to be multiplied. There is 1 mark for this.

Multiply the total surface area of cuboid **S** by 25 to find the total surface area of cuboid **T**. 1 mark for the correct answer.

Cube the scale factor to find the number by which the volume has been multiplied. There is 1 mark for this.

Divide the volume of cuboid **T** by 125 to find the volume of cuboid **S**. 1 mark for the correct answer.

WARNING ⚠

A common error is to multiply the volume by 125, instead of dividing it by 125. This is not sensible, as cuboid **S** is smaller than cuboid **T** and so it must have a smaller volume.

Two cylinders, **P** and **Q**, are geometrically similar.

P   Q

The volume of cylinder **P** is 5 cm³ and the volume of cylinder **Q** is 320 cm³.

The total surface area of cylinder **P** is 15 cm². Calculate the total surface area of cylinder **Q**.

$\frac{Vol\ Q}{Vol\ P}$ = $\frac{320}{5}$ = 64

k^3 = 64

$k = \sqrt[3]{64}$ = 4

Multiply area by 4^2 = 16
Area Q = 15 cm² × 16
        = 240 cm²

Work out $\frac{\text{volume of } \mathbf{Q}}{\text{volume of } \mathbf{P}}$ to find the number by which the volume has been multiplied. There is 1 mark for this.

This number is (scale factor)³.

Find the scale factor, which is the cube root of this number. There is 1 mark for this.

Square the scale factor to find the number by which the total surface area must be multiplied.

1 mark for the correct answer.

Multiply the total surface area of cylinder *P* by 16 to find the total surface area of cylinder *Q*. There is 1 mark for this.

1 Quadrilaterals **P** and **Q** are similar.
The area of quadrilateral **P** is 10 cm².

Calculate the area of quadrilateral **Q**.

2 Pentagons **P** and **Q** are similar.
The area of pentagon **Q** is 16 times the area of pentagon **P**.

Calculate the value of **a** x,
b y.

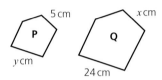

3 Triangles **P** and **Q** are similar.
The area of triangle **P** is 40 cm² and the area of triangle **Q** is 250 cm².

Calculate the value of **a** x,
b y.

4 Pyramids **S** and **T** are similar.
   a The surface area of pyramid **T** is 600 cm².
   Calculate the surface area of pyramid **S**.
   b The volume of pyramid **S** is 48 cm³.
   Calculate the volume of pyramid **T**.

5 The two containers **A** and **B** are similar.
The surface area of container **A** is 1000 cm².
The surface area of container **B** is 62.5 cm².
The volume of container **A** is 2500 cm³.
Calculate the volume of container **B**.

Diagram **NOT** accurately drawn

(1384 November 1994)

6 Two prisms, **A** and **B**, are mathematically similar.
The volume of prism **A** is 12 000 cm³.
The volume of prism **B** is 49 152 cm³.
The total surface area of prism **B** is 9728 cm².
Calculate the total surface area of prism **A**.

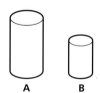

Diagram **NOT** accurately drawn

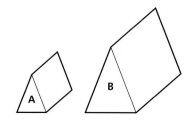

(1387 November 2006)

7 The diagram shows a model.
The model is a cuboid with a pyramid on top.
The base of the model is a square with sides of length 5 cm.
The height of the cuboid in the model is 10 cm.
The height of the pyramid in the model is 6 cm.

a Calculate the volume of the model.

The model represents a concrete post.
The model is built to a scale of 1 : 30
The surface area of the model is 290 cm³.

b Calculate the surface area of the post.
Give your answer in square metres.

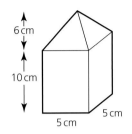

Diagram **NOT**
accurately drawn

REFERENCE
For a reminder of how
to find the volume of a
pyramid, turn to page 84.

HINT

1 m³ = 1 000 000 cm³

(1387 November 2005)

8 On a farm, wheat grain is stored in a cylindrical tank.
The cylindrical tank has an internal diameter of
6 metres and a height of 9 metres.

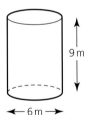

Diagram **NOT**
accurately drawn

9 m

6 m

a Calculate the volume, in m³, of the tank.
Give your answer correct to 2 decimal places.

1 m³ of wheat weighs 0.766 tonnes.

b Calculate the weight, in tonnes, of wheat grain in the
storage tank when it is full.
Give your answer correct to the nearest tonne.

A scale model is to be made of the farm.
The scale of the model is 1 : 50
The water tank on the real farm has a volume of 120 m³.

c Calculate the volume, in cm³, of the similar water tank in this
scale model.

(1384 November 1996)

19 Arc length and sector area

KEY FACTS

For a sector with an angle of $x°$ at the centre of a circle of radius r

- Arc length = $\frac{x}{360} \times 2\pi r$

- Sector area = $\frac{x}{360} \times \pi r^2$

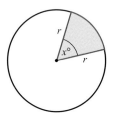

EXAM FACTS

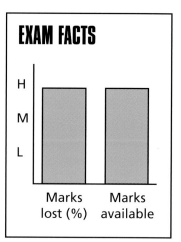

Getting it right

The diagram represents the landing area for a shot put competition.

$OACB$ is a sector of a circle, centre O, radius 10 m.
Angle $AOB = 80°$.

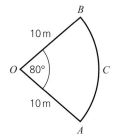

Diagram **NOT** accurately drawn

EXAM TIP

Diagram NOT accurately drawn means that taking measurements from the diagram will give WRONG answers.

a Calculate the length of the arc ACB.
Give your answer correct to 3 significant figures.
b Calculate the area of the sector $OACB$.
Give your answer correct to 3 significant figures.

(1384 June 1994)

a $\frac{80}{360} \times 2 \times \pi \times 10$

$= 13.962...$

Length of arc ACB = 14.0 cm (to 3 s.f.)

$\frac{80}{360}$ is the fraction of a whole circle. There is no need to simplify this fraction.

$2 \times \pi \times 10$ is the circumference of the whole circle. $\frac{80}{360} \times 2 \times \pi \times 10$ scores 1 mark, even if it is not evaluated correctly.

b $\frac{80}{360} \times \pi \times 10^2$

$= 69.813...$

Area of sector OACB = 69.8 cm² (to 3 s.f.)

Write down at least 4 figures of the calculator display.

Round the arc length to 3 significant figures. The zero in 14.0 **is** necessary.

$\pi \times 10^2$ is the area of the whole circle. $\frac{80}{360} \times \pi \times 10^2$ scores 1 mark, even if it is not evaluated correctly.

Write down at least 4 figures of the calculator display, THEN round the sector area to 3 significant figures.

Now try these

If your calculator does not have a π button, take the value of π to be 3.142

1 Calculate **a** the arc length and **b** the area of each of these sectors.
Give your answers correct to 3 significant figures.

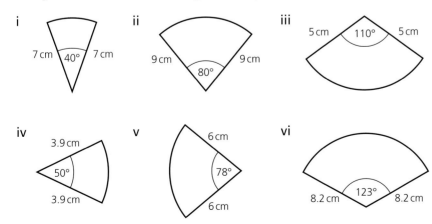

i 7 cm 40° 7 cm

ii 9 cm 80° 9 cm

iii 5 cm 110° 5 cm

iv 3.9 cm 50° 3.9 cm

v 6 cm 78° 6 cm

vi 8.2 cm 123° 8.2 cm

2 The diagram shows the shape *PQRST*.
RST is a circular arc with centre *P* and radius 18 cm.
Angle *RPT* = 40°.

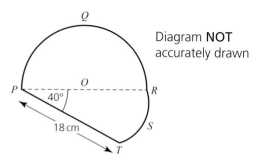

Diagram **NOT**
accurately drawn

a Calculate the length of the circular arc *RST*.
Give your answer correct to 3 significant figures.

PQR is a semicircle with centre *O*.
b Calculate the **total** area of the shape *PQRST*.
Give your answer correct to 3 significant figures.

(1385 November 2000)

3 The diagram shows a prism.
It has a cross-section in the shape of a sector *AOB* of a circle, centre *O*.
The radius of the sector is 9 cm and the
length of the prism is 15 cm.
Angle *AOB* = 54°.

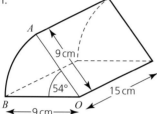

Diagram **NOT**
accurately drawn

a Calculate the area of sector *AOB*.
Give your answer correct to 3 significant figures.
b Calculate the total surface area of the prism.
Give your answer correct to 3 significant figures.

HINT

The total surface area consists
of 2 sectors and 3 rectangles.
(The arc length *AB* is the side
of one of the rectangles.)

4 The diagram shows a sector *ABC* of a circle with centre *O*.
OA = *OC* = 10.4 cm.
Angle *AOC* = 120°.

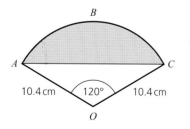

Diagram **NOT**
accurately drawn

HINT

Use '$A = \frac{1}{2}ab\sin C$' to find
the area of triangle *AOC*.

REFERENCE

For a reminder of
Area of triangle = $\frac{1}{2}ab\sin C$
see pages 89 and 90.

a Calculate the length of the arc *ABC* of the sector.
Give your answer correct to 3 significant figures.
b Calculate the area of the shaded segment *ABC*.
Give your answer correct to 3 significant figures.

(1387 June 2006)

5 The diagram shows an isosceles triangle ABC.
$AB = AC = 12$ cm.
P is the midpoint of AB.
Q is the midpoint of AC.
Angle $PAQ = 50°$.
APQ is a sector of a circle, centre A.

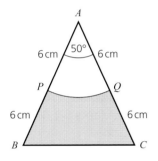

Diagram **NOT** accurately drawn

HINT

use '$A = \frac{1}{2}ab\sin C$' to find the area of triangle ABC.

REFERENCE

For a reminder of Area of triangle = $\frac{1}{2}ab\sin C$ see pages 89 and 90.

Calculate the area of the shaded region.
Give your answer correct to 3 significant figures.

(1388 November 2006)

6 The diagram shows a sector of a circle, centre O, radius 10 cm.
Calculate the area of the sector.

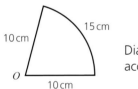

Diagram **NOT** accurately drawn

(1388 March 2005)

SKILL

Use Pythagoras' theorem and trigonometry to solve complex problems in two dimensions

EXAM FACTS

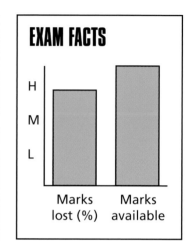

Marks lost (%)   Marks available

KEY FACTS

- Pythagoras' theorem $c^2 = a^2 + b^2$

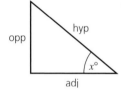

- Trigonometry

$$\sin x° = \frac{\text{opp}}{\text{hyp}}, \text{opp} = \text{hyp} \times \sin x°$$

$$\cos x° = \frac{\text{adj}}{\text{hyp}}, \text{adj} = \text{hyp} \times \cos x°$$

$$\tan x° = \frac{\text{opp}}{\text{adj}}, \text{opp} = \text{adj} \times \tan x°$$

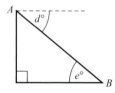

- The angle $e°$ is the angle of elevation of A from B.

  The angle $d°$ is the angle of depression of B from A.

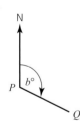

- The angle $b°$ is the bearing of Q from P. A bearing is measured clockwise from north (N).

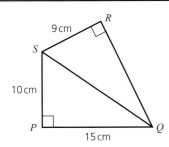

PQRS is a quadrilateral.
Angle SPQ = angle SRQ = 90°.
PQ = 15 cm, PS = 10 cm and SR = 9 cm.
Work out the size of angle RQS.
Give your answer correct to 1 decimal place.

In triangle PQS, Pythagoras' theorem gives

$SQ^2 = 10^2 + 15^2$

$SQ^2 = 100 + 225 = 325$

$SQ = \sqrt{325} = 18.027756...$

$\sin RQS = \dfrac{9}{SQ} = \dfrac{9}{18.027756...}$

$\sin RQS = 0.49923...$

Angle $RQS = 29.949...°$

Angle $RQS = 29.9°$ (to 1 d.p.)

The length of one other side in triangle QRS has to be known before trigonometry can be used to work out the size of angle RQS. Pythagoras' theorem can be used in triangle PQS to work out the length of hypotenuse SQ, which is also the hypotenuse of triangle QRS.

Write down Pythagoras' theorem in triangle PQS. The hypotenuse is SQ.

WARNING

The length of SQ is to be used in triangle QRS. To avoid losing marks due to premature approximation, do not round the length of SQ. Use at least 4 significant figures or even all the figures in the calculator display.

If SQ had been incorrectly calculated, writing this with the incorrect length of SQ would still gain a mark.

In triangle QRS, side SR is opposite angle RQS and SQ is the hypotenuse. This means that 'sine' has to be used.

EXAM TIP

Always write down an unrounded value first before giving your final answer correct to 1 decimal place.

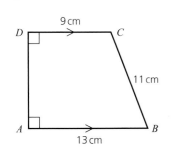

Here is a trapezium $ABCD$.

Work out the area of the trapezium. Give your answer correct to 3 significant figures.

The area of a trapezium is
$$A = \tfrac{1}{2}(a + b)h$$
In this question a = 9 cm, b = 13 cm and h has to be calculated.

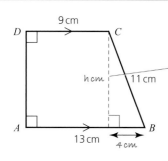

9 cm

D ⌐ ———→ C

h cm | 11 cm

A ⌐ ———→ | ⌐ B
13 cm   4 cm

$11^2 = 4^2 + h^2$

$121 = 16 + h^2$

$h^2 = 121 - 16 = 105$

$h = \sqrt{105} = 10.24695...$

Area of trapezium $= \frac{1}{2}(9 + 13) \times 10.24695...$

Area $= 112.716...$

Area $= 113$ cm² (to 3 s.f.)

Drawing this line produces a right angled triangle with hypotenuse of length 11 cm.

This side has length 4 cm as $13 - 9 = 4$

Write down Pythagoras' theorem in this triangle. The hypotenuse has length 11 cm.

h is the distance between the parallel sides in the trapezium.

WARNING

Write down and use all the figures in the calculator display to avoid errors due to premature approximation.

Write down an uncorrected value before correcting your answer to 3 significant figures.

Now try these

1 $PQRS$ is a trapezium.
   PQ is parallel to SR.
   Angle SPQ = angle PSR = 90°.
   Angle QRS = 60°.
   PQ = 14 cm.
   SR = 20 cm.

   Work out the area of the trapezium.
   Give your answer correct to 3 significant figures.

   Diagram **NOT** accurately drawn

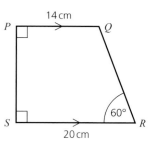

   (1388 March 2007)

2 ABC is a triangle
   N is the point on AB such that angle ANC = 90°.
   AN = 8 cm, AC = 11 cm and NB = 12 cm.

   a Work out the length of NC.
      Give your answer correct to 3 significant figures.
   b Work out the size of angle ABC.
      Give your answer correct to 3 significant figures.

   Diagram **NOT** accurately drawn

3 $AC = 16$ cm
Angle $ABC = 90°$
Angle $CAB = 30°$
$BC = BD$
$CD = 12$ cm

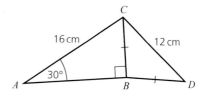

Diagram **NOT**
accurately drawn

Calculate the area of triangle BCD.
Give your answer correct to 3 significant figures.

(1388 January 2004)

4 The diagram shows a circle of radius 4 cm
inside a square $ABCD$ of side 8 cm.
P is a point of intersection of the circle and the
diagonal AC of the square.
Show that $AP = 1.66$ cm, correct to 3 significant figures.

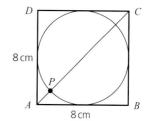

Diagram **NOT**
accurately drawn

(4400 November 2005)

5 The diagram represents a vertical flagpole, AB. The flagpole is supported
by two ropes, BC and BD, fixed to the horizontal ground at C and at D.

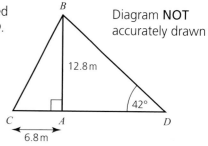

Diagram **NOT**
accurately drawn

$AB = 12.8$ m. $AC = 6.8$ m.
Angle $BDA = 42°$.

a Calculate the size of angle BCA.
Give your answer correct to 3 significant figures.
b Calculate the length of the rope BD.
Give your answer correct to 3 significant figures.

(1387 November 2003)

6 The diagram is part of a map showing the positions of
three Nigerian towns.
Kaduna is due North of Aba.

a Calculate the direct distance between Lagos and Kaduna.
Give your answer to the nearest kilometre.
b Calculate the distance between Kaduna and Aba.
Give your answer to the nearest kilometre.

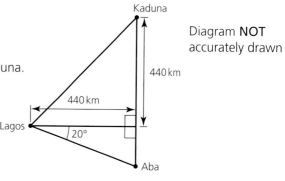

Diagram **NOT**
accurately drawn

(1384 June 1994)

7 The diagram represents the frame for part of a building.
BC and CD are equal in length.
BD and AE are horizontal.

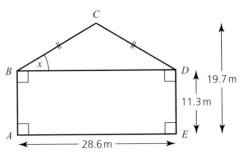

Diagram **NOT** accurately drawn

a Calculate the length BC.
Give your answer correct to 3 significant figures.

b Calculate the size of the angle marked x.
Give your answer in degrees correct to 1 decimal place.

(1385 November 1998)

8 A boy starts at A. He walks 2x metres to B,
up a slope of 15.2° to the horizontal.
He then walks x metres from B to C, down a slope
of 20.3° to the horizontal.
C is 88 metres above the level of A.

Calculate the value of x.
Give your answer correct to 3 significant figures.

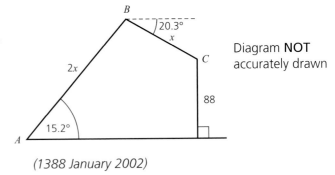

Diagram **NOT** accurately drawn

(1388 January 2002)

9 ABCDE is a pentagon.
BC = ED = 6 m.
Angle BCD = angle CDE = 90°.
Angle BAE = 56°.

The point F lies on CD so that AF is the line of
symmetry of the pentagon and AF = 10 m.

Calculate the perimeter of the pentagon.
Give your answer correct to 3 significant figures.

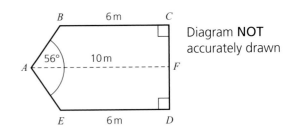

Diagram **NOT** accurately drawn

(1388 March 2006)

10 PQ is a vertical tower of height 20 m.
M and N are points on horizontal ground so that MQN is a straight line.
MN = 45 m.

The angle of elevation of P from M is 50°.

Calculate the angle of elevation of P from N.
Give your answer to the nearest degree.

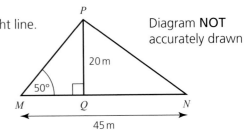

Diagram **NOT** accurately drawn

21 Pythagoras' theorem and trigonometry in 3-D

SKILL

Use Pythagoras' theorem and trigonometry in problems in three dimensions

KEY FACTS

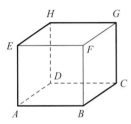

- Problems with cuboids and other 3-D shapes involve identifying right-angled triangles and applying Pythagoras' theorem and trigonometry to them.

  In the diagram, examples of right-angled triangles are

  Triangle *ACG*

  Triangle *EHD*

  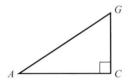

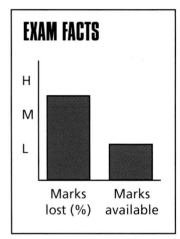

  AC is the projection of *AG* on the plane *ABCD* so that angle *GAC* is the angle between *AG* and the plane *ABCD*.

EXAM FACTS

Marks lost (%) — Marks available

REFERENCE

For a reminder of how to use Pythagoras' theorem and trigonometry in 2-D, turn to pages 74 to 76.

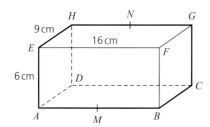

$ABCDEFGH$ is a cuboid with length 16 cm, width 9 cm and height 6 cm. The midpoint of AB is M and the midpoint of HG is N.

Calculate the length of i MN, ii MG.

Give your answers correct to 3 significant figures.

i

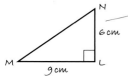

$MN^2 = 9^2 + 6^2$
$MN^2 = 81 + 36 = 117$
$MN = \sqrt{117} = 10.81665\ldots$
$MN = 10.8\,\text{cm}$ (to 3 s.f.)

ii

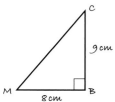
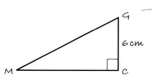

In triangle BCM, Pythagoras' theorem gives
$MC^2 = 8^2 + 9^2 = 64 + 81$
$MC^2 = 145$

In triangle CGM, Pythagoras' theorem gives
$MG^2 = MC^2 + 6^2 = 145 + 36$
$MG^2 = 181$

$MG = \sqrt{181} = 13.4536\ldots$
$MG = 13.5\,\text{cm}$ (to 3 s.f.)

Look for a right-angled triangle with MN as one side. With L as the midpoint of DC, triangle LMN is a suitable triangle as $LM = 9$ cm and $LN = 6$ cm.

Write down Pythagoras' theorem for triangle LMN.

EXAM TIP

Write down at least 4 figures before giving your answer correct to 3 significant figures.

Triangle CGM is right-angled with MG as the hypotenuse and $CG = 6$ cm. To work out the length of MC, right-angled triangle BCM has to be used.

It is not necessary to write down the length of MC. Pythagoras' theorem is to be used in triangle CGM so MC^2 is needed not MC.

EXAM TIP

Write down at least 4 figures before giving your answer correct to 3 significant figures.

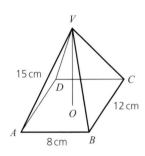

The diagram shows a rectangular-based pyramid. The base of the pyramid is $ABCD$, where $AB = 8$ cm and $BC = 12$ cm.

The vertex of the pyramid is V and O is the centre of the base so that VO is perpendicular to the base of the pyramid.

$AV = BV = CV = DV = 15$ cm

a   Show that, correct to 3 significant figures, $AO = 7.21$ cm.
b   Calculate the height of the pyramid.
    Give your answer correct to 3 significant figures.

a   **By Pythagoras' theorem**

$AC^2 = 8^2 + 12^2 = 64 + 144 = 208$

$AC = \sqrt{208} = 14.4222051$

$AO = \frac{1}{2} \times 14.4222051 = 7.2111...$

$AO = 7.21$ cm (to 3 s.f.)

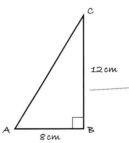

$AO = \frac{1}{2}AC$ so use Pythagoras' theorem in triangle ABC.

b   **By Pythagoras' theorem**

$15^2 = AO^2 + OV^2$

$OV^2 = 15^2 - AO^2$

$OV^2 = 15^2 - 7.2111...^2$

$OV^2 = 225 - 52 = 173$

$OV = \sqrt{173} = 13.1529...$

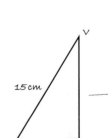

The height of the pyramid is the length of OV. Use Pythagoras' theorem in triangle OAV.

The height of the pyramid = 13.2 cm (to 3 s.f.)

1 The diagram shows a cuboid.
A, *B*, *C*, *D* and *E* are five vertices of the cuboid.
AB = 5 cm, *BC* = 8 cm, *CE* = 3 cm.

Calculate the length of
a *BE*, **b** *DE*, **c** *AE*.

Give your answers correct to 3 significant figures.

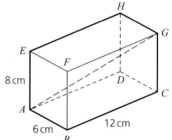

Diagram **NOT** accurately drawn

2 The diagram represents a cuboid *ABCDEFGH*.
AB = 6 cm, *BC* = 12 cm, *AE* = 8 cm.

  a Calculate the length of *AG*.
     Give your answer correct to 3 significant figures.
  b Calculate the size of the angle between *AG* and
     the plane *ABCD*.
     Give your answer correct to 1 decimal place.

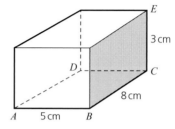

Diagram **NOT** accurately drawn

3 The diagram shows a cylinder with a height of 10 cm and
a radius of 4 cm.
The length of a pencil is 13 cm. The pencil cannot be broken.
Show that this pencil cannot fit inside the cylinder.

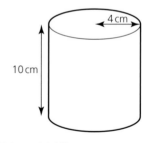

Diagram **NOT** accurately drawn

(1387 June 2003)

4 The diagram represents a cone.
The height of the cone is 12 cm.
The diameter of the base of the cone is 10 cm.
The vertex of the cone is *V* and *AB* is a diameter of the base.

  a Calculate the slant height, *VA*, of the cone.
  b Calculate the size of the angle between *VA* and the
     base of the cone.
     Give your answer correct to the nearest 0.1°.

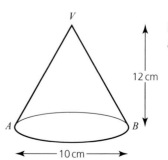

Diagram **NOT** accurately drawn

5 *ABCDEF* is a triangular prism.
The rectangular plane *BCFE* is horizontal.
The rectangular plane *ABED* is vertical.
Triangle *ABC* is right-angled at B.

AB = 12 cm, *BC* = 5 cm, *CF* = 10 cm.

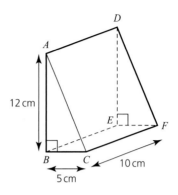

Diagram **NOT**
accurately drawn

a Calculate the length of
 i *BF*, **ii** *AF*, **iii** *AE*
 Give your answers correct to 3 significant figures.
b Calculate the size of the angle between *AF* and the horizontal.
 Give your answer correct to the nearest degree.

6 The diagram shows a rectangular-based pyramid. The base of the
pyramid is *ABCD*, where *AB* = 10 cm and *BC* = 15 cm. The vertex
of the pyramid is *V* and *O* is the centre of the base so that *VO* is
perpendicular to the base of the pyramid.
The height, *VO*, of the pyramid is 20 cm.

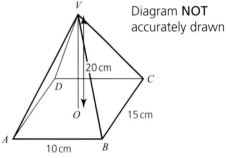

Diagram **NOT**
accurately drawn

a Work out the length of
 i *AO*, **ii** *VA*.

The midpoint of *AB* is *N*.
b Calculate the length of *VN*.
c Calculate the size of the angle between *VN* and the base, *ABCD*,
 of the pyramid.

Give your answers correct to 3 significant figures.

7 *ABCD* is a horizontal rectangular field.
AB = 50 m, *BC* = 27 m.
AT is a vertical mast.

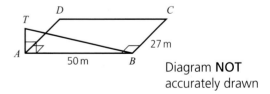

a The angle of elevation of *T* from *B* is 19°
 Calculate the length of *AT*.
 Give your answer correct to 3 significant figures.
b Calculate the distance from *C* to *T*.
 Give your answer correct to 3 significant figures.

Diagram **NOT**
accurately drawn

(4400 May 2005)

REFERENCE

For a reminder of angles of
elevation and depression, see
page 74.

SKILL

Find the volume and surface area of a variety of complex shapes

EXAM FACTS

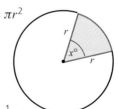

Marks lost (%)   Marks available

KEY FACTS

This information is provided on the formulae page of the GCSE examination paper.

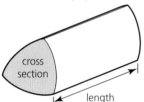

cross section

length

Volume of a prism
= area of cross section × length

Volume of sphere = $\frac{4}{3}\pi r^3$

Surface area of sphere = $4\pi r^2$

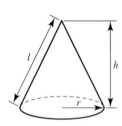

Volume of cone = $\frac{1}{3}\pi r^2 h$

Curved surface area of sphere = $\pi r l$

In addition, you should know

- Area of a rectangle = length × width $A = lw$

- Area of a triangle = $\frac{1}{2}$ × base × height $A = \frac{1}{2}bh$

- Area of a trapezium = $\frac{1}{2}$ × sum of parallel sides × distance between them
  $A = \frac{1}{2}(a + b)h$

- Circumference of a circle = π × diameter = 2 × π × radius
  $C = \pi d = 2\pi r$

- Area of a circle = π × (radius)^2 $A = \pi r^2$

- Arc length = $\frac{x}{360}$ × $2\pi r$

- Sector area = $\frac{x}{360}$ × πr^2

- Volume of a pyramid = $\frac{1}{3}$ × base area × height $V = \frac{1}{3}Ah$

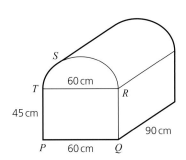

Diagram **NOT** accurately drawn

The diagram shows a prism of length 90 cm.
The cross section, *PQRST*, of the prism is a semicircle above a rectangle.
PQRT is a rectangle. *RST* is a semicircle with diameter *RT*.

$PQ = RT = 60$ cm. $PT = QR = 45$ cm.

Calculate the volume of the prism.
Give your answer correct to 3 significant figures.
State the units of your answer.

(1387 November 2006)

Area of rectangle PQRT $= 60 \times 45$
$= 2700 \ cm^2$

Area of semicircle RST $= \dfrac{\pi \times 30^2}{2}$
$= 1413.7... \ cm^2$

Area of cross section PQRST $= 2700 + 1413.7$
$= 4113.7... \ cm^2$

Volume of the prism $= 4113.7... \times 90$
$= 370234.5...$

$= 370\,000 \ cm^3$ (to 3 s.f.)

To work out the area of the rectangle, multiply its length by its width.
This scores 1 mark.

Show how to work out the area of the semicircle. This scores 1 mark.

Write down at least 4 figures.

WARNING

A common error is to use 60 cm as the radius of the semicircle.

Add the two results to find the area of cross section of the prism.

Multiply the area of cross section of the prism by its length. This scores 1 mark.

1 mark for the correct answer.

EXAM TIP

You will gain a mark for stating the correct units, even if everything else is wrong.

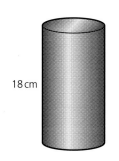 

Diagram **NOT**
accurately drawn

18 cm

EXAM TIP

The radius of a cylinder is equal to the radius of a sphere.
The length of the cylinder is 18 cm.
The **total** surface area of the cylinder is twice the surface area of the sphere.

Work out the volume of the sphere.
Give your answer as a multiple of π.

Questions on the non-calculator paper will sometimes ask you to leave your answer in terms of π.

Let r cm be the radius of the cylinder and of the sphere.

Total surface area of cylinder $= 2\pi r \times 18 + 2\pi r^2$
$= 36\pi r + 2\pi r^2$

Surface area of sphere $= 4\pi r^2$

Total surface area of cylinder $= 2 \times$ Surface area of sphere

So $36\pi r + 2\pi r^2 = 2 \times 4\pi r^2$
$36\pi r + 2\pi r^2 = 8\pi r^2$
$36\pi r = 6\pi r^2$
$36 = 6r$
$r = 6$

Volume of sphere $= \frac{4}{3} \times \pi \times 6^3 = 288\pi$ cm³

Substitute $h = 18$ into the formula for the total surface area, A, of a solid cylinder of radius r and length h
$A = 2\pi rh + 2\pi r^2$

Divide both sides by π and by r. (When both sides are divided by r, the solution $r = 0$ is lost, but only the positive solution is relevant to the answer.)

Substitute $r = 6$ into the formula for the surface area, A, of a sphere of radius r
$A = \frac{4}{3}\pi r^3$

Now try these

1 The diagram shows a water tank.
The tank is a hollow cylinder joined to a hollow hemisphere at the top.
The tank has a circular base.
Both the cylinder and the hemisphere have a diameter of 46 cm.
The height of the tank is 90 cm.

Work out the volume of water which the tank holds when it is full.
Give your answer correct to 3 significant figures.

(1385 June 2000)

Diagram **NOT**
accurately drawn

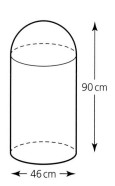

90 cm

← 46 cm →

2 The diagram shows a solid shape made from a hemisphere and a cone.
The hemisphere has a radius of 5 cm.
The cone has a radius of 5 cm, a vertical height of 12 cm and a slant height of 13 cm.

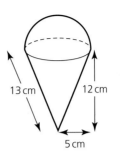

Diagram **NOT** accurately drawn

13 cm   12 cm

5 cm

  a Calculate the volume of the solid.
   Give your answer correct to 3 significant figures.
   State the units of your answer.
  b Calculate the total surface area of the solid.
   Give your answer correct to 3 significant figures.
   State the units of your answer.

3 A tent has a groundsheet as its horizontal base.
The shape of the tent is a triangular prism of length 8 metres, with two identical half right-circular cones, one at each end.
The vertical cross section of the prism is an isosceles triangle of height 2.4 metres and base 3.6 metres.

Diagram **NOT** accurately drawn

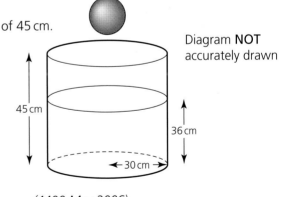

2.4 m   3.6 m   8 m

TENT

  a Calculate the area of the groundsheet.
   Give your answer in m², correct to 1 decimal place.
  b Calculate the total volume of the tent.
   Give your answer in m³, correct to 1 decimal place.

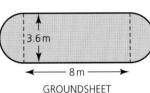

3.6 m   8 m

GROUNDSHEET

(1385 June 1999)

4 A cylindrical tank has a radius of 30 cm and a height of 45 cm.
The tank contains water to a depth of 36 cm.
A metal sphere is dropped into the water and is completely covered.
The water level rises by 5 cm.

Calculate the radius of the sphere.

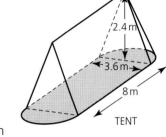

Diagram **NOT** accurately drawn

45 cm   36 cm   30 cm

(4400 May 2006)

5 The radius of the base of a cone is 5.7 cm.
Its slant height is 12.6 cm.

Calculate the volume of the cone.

Give your answer correct to 3 significant figures.

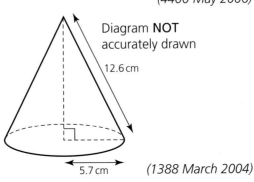

Diagram **NOT** accurately drawn

12.6 cm   5.7 cm

HINT
Use Pythagoras' theorem to find the vertical height.

(1388 March 2004)

6 The diagram represents a cone.
The height of the cone is 12 cm.
The diameter of the base of the cone is 10 cm.

Calculate the curved surface area of the cone.
Give your answer as a multiple of π.

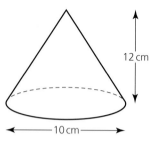

Diagram **NOT**
accurately drawn

12 cm

10 cm

(1388 January 2004)

7 A greenhouse consists of a pyramid on top of a prism.
The cross section of the prism and the base of the pyramid
is a regular octagon. Each side of the octagon is 0.80 m long.
The height of the prism is 1.73 m.
The height of the pyramid is 0.68 m.

Calculate the volume of the greenhouse.
Give your answer correct to 3 significant figures.

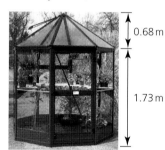

Diagram **NOT**
accurately drawn

0.68 m

1.73 m

0.80 m

(1385 June 2002)

8 The diagram shows a sector OAB of a circle, centre O.
Angle $AOB = 171°$.
OA and OB are joined to make a cone.

Calculate the vertical height, in centimetres, of the cone.
Give your answer correct to 3 significant figures.

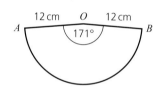

12 cm O 12 cm

A 171° B

Diagram **NOT**
accurately drawn

(1385 November 2002)

9 The radius of a sphere is 3 cm.
The radius of the base of a cone is also 3 cm.
The volume of the sphere is 3 times the volume
of the cone.

Work out the curved surface area of
the cone.
Give your answer as a multiple of π.

Diagram **NOT**
accurately drawn

3 cm

3 cm

(1387 November 2003)

REFERENCE

See pages 42 and 43 for a
reminder of how to factorise
quadratic expressions.

10 a Factorise $2x^2 + 19x - 33$

A cone fits exactly on top of a hemisphere to form a solid toy.
The radius, CA, of the base of the cone is 3 cm.

b Show that the **total** surface area of the toy is 33π cm².

The radius of the base of a cylinder is x cm.
The height of the cylinder is 9.5 cm longer than the radius
of its base.
The area of the **curved** surface of the cylinder is equal to
the **total** surface area, 33π cm², of the toy.

c Calculate the height of the cylinder.

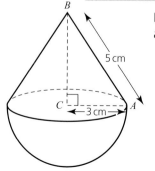

B

5 cm

C 3 cm A

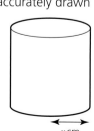

Diagram **NOT**
accurately drawn

x cm

(1385 November 2000)

23 Using the sine rule, the cosine rule and $\frac{1}{2}ab \sin C$

SKILL

Use the sine rule, the cosine rule and area of triangle = $\frac{1}{2}ab \sin C$ to solve problems involving scalene triangles

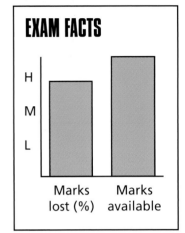
KEY FACTS

In any triangle ABC

(The highlighted formulae are given on the examination paper's formulae page.)

- **Sine rule** (Used when two sides and a non-included angle or two angles and a side are known)

$$\frac{a}{\sin A} = \frac{b}{\sin B} = \frac{c}{\sin C}$$

$$\frac{\sin A}{a} = \frac{\sin B}{b} = \frac{\sin C}{c}$$

- **Cosine rule** (Used when two sides and the included angle or all three sides are known)

$$a^2 = b^2 + c^2 - 2bc \cos A$$

$$\cos A = \frac{b^2 + c^2 - a^2}{2bc}$$

- **Area of triangle** = $\frac{1}{2}ab \sin C$

Getting it right

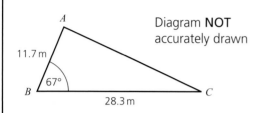

Diagram **NOT** accurately drawn

$AB = 11.7$ m, $BC = 28.3$ m, Angle $ABC = 67°$.

a Calculate the area of the triangle *ABC*.
   Give your answer correct to 3 significant figures.
b Calculate the length of *AC*.
   Give your answer correct to 3 significant figures.

(1387 November 2003)

Applying '$\frac{1}{2}ab \sin C$' will score 1 mark.

EXAM TIP

When using any trigonometry formulae, make sure that your calculator is set in degree mode.

a Area of triangle $ABC = \frac{1}{2} \times 11.7 \times 28.3 \times \sin 67°$
   $= 152.39.....$

   Area of triangle $= 152$ m² (to 3 s.f.)

b $AC^2 = 11.7^2 + 28.3^2 - 2 \times 11.7 \times 28.3 \times \cos 67°$
   $= 937.78 - 258.7499...$
   $= 679.030....$
   $AC = \sqrt{679.03} = 26.058...$
   $AC = 26.1$ m (to 3 s.f.)

Use the cosine rule because 2 sides and the included angle are known.

Correct substitution into the cosine rule will score 1 mark.

EXAM TIP

Write down at least 4 figures from your calculator display and remember to round your answer to 3 significant figures.

WARNING

A common mistake here is to evaluate AC^2 as
$(11.7^2 + 28.3^2 - 2 \times 11.7 \times 28.3) \cos 67°$
that is $275.56 \times \cos 67°$.
The correct order of evaluation will score 1 mark.

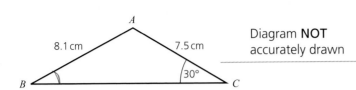

In triangle ABC, $AB = 8.1$ cm, $AC = 7.5$ cm, angle $ACB = 30°$.
Calculate the size of angle ABC.
Give your answer correct to 3 significant figures.

$$\frac{\sin B}{7.5} = \frac{\sin 30°}{8.1}$$

$$\sin B = \frac{7.5 \times \sin 30°}{8.1} = 0.46296...$$

$$B = 27.578...°$$

Angle $ABC = 27.6°$ (to 3 s.f.)

Making $\sin B$ the subject will score 1 mark.

Two sides and a non-included angle are known so use the sine rule. As the size of an angle has to be calculated use
$$\frac{\sin A}{a} = \frac{\sin B}{b} = \frac{\sin C}{c}$$

Applying the sine rule to this triangle will score 1 mark.

EXAM TIP

Write down at least 4 figures from your calculator display and remember to give your answer to 3 significant figures.

EXAM TIP

Make sure your answer is sensible.
As $7.5 < 8.1$ angle ABC must be $< 30°$.

Now try these

1 ABC is a triangle.
$AB = 8$ cm. $BC = 14$ cm. Angle $ABC = 106°$.
Calculate the area of the triangle.
Give your answer correct to 3 significant figures.

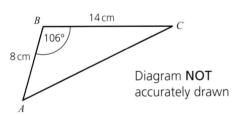

Diagram **NOT** accurately drawn

(1387 June 2005)

2 In triangle ABC, $AC = 7$ cm, $BC = 10$ cm, angle $ACB = 73°$.
Calculate the length of AB.
Give your answer correct to 3 significant figures.

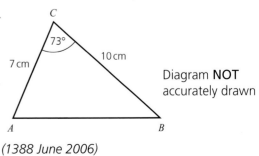

Diagram **NOT** accurately drawn

(1388 June 2006)

3 $BC = 9.4$ cm. Angle $BAC = 123°$. Angle $ABC = 35°$

  a Calculate the length of AC.
    Give your answer correct to 3 significant figures.
  b Calculate the area of triangle ABC.
    Give your answer correct to 3 significant figures.

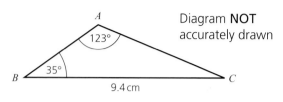

Diagram **NOT** accurately drawn

(4400 May 2005)

4 The lengths of the sides of a triangle are 4.2 cm, 5.3 cm and 7.6 cm.

  a Calculate the size of the largest angle of the triangle.
    Give your answer correct to 1 decimal place.
  b Calculate the area of the triangle.
    Give your answer correct to 3 significant figures.

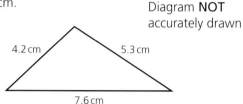

Diagram **NOT** accurately drawn

(1387 November 2006)

5 Three posts, A, B and C are on a horizontal school field.
  AC is 41.3 m and C is due east of A.
  The bearing of B from A is 036°.
  The bearing of B from C is 308°.
  Calculate the perimeter of triangle ABC.
  Give your answer in metres, correct to the nearest metre.

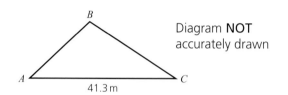

Diagram **NOT** accurately drawn

REFERENCE

For a reminder of bearings see page 84.

6 The diagram shows a triangle ABC.
  $AB = 6$ cm, $BC = 12$ cm, angle $ABC = 67°$.

  a Calculate the length of AC.
    Give your answer correct to 3 significant figures.
  b Calculate the size of angle ACB.
    Give your answer correct to 3 significant figures.

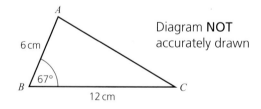

Diagram **NOT** accurately drawn

7 Calculate the area of the triangle.
  Give your answer correct to 3 significant figures.

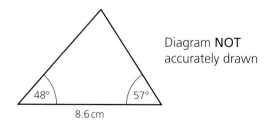

Diagram **NOT** accurately drawn

(4400 November 2006)

8 The diagram shows a quadrilateral *ABCD*.
   AB = 8.3 cm, *BC* = 7.8 cm, *CD* = 5.4 cm and *AD* = 6.1 cm.
   Angle *BAD* = 71°.

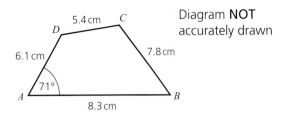

Diagram **NOT** accurately drawn

   a Calculate the area of triangle *ABD*.
      Give your answer correct to 3 significant figures.
   b Calculate the size of angle *BCD*.
      Give your answer correct to 1 decimal place.

9 The 12-sided window is made up of squares and equilateral triangles.
   The perimeter of the window is 15.6 m.

   Calculate the area of the window.
   Give your answer correct to 3 significant figures.

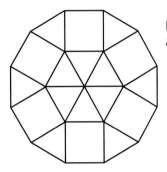

Diagram **NOT** accurately drawn

(1388 March 2005)

10 The diagram shows a lighthouse *L* and two points *A* and *B* on the sea.
    AL = 5 km and *BL* = 6 km.
    Angle *ALB* = 150°.

    A boat sails in a straight line from point *A* to point *B*.
    a Calculate the distance *AB*.
       Give your answer in kilometres, correct to
       3 significant figures.

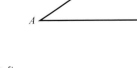

Diagram **NOT** accurately drawn

    At its shortest distance from the lighthouse,
    the boat is at the point *X* on the line *AB*.

    b Calculate the distance *LX*.
       Give your answer in kilometres, correct to 3 significant figures.

(1384 June 1999)

REFERENCE

Questions using $\frac{1}{2}ab \sin C$ together with finding the area of a sector can be found on pages 72 and 73.

EXAM FACTS

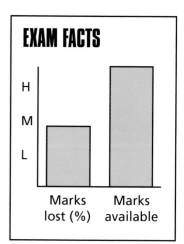

Marks lost (%) Marks available

KEY FACTS

- The angle subtended by an arc at the centre of a circle is twice the angle subtended at the circumference.

  $b = 2a$

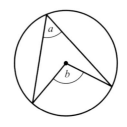

- The angle in a semicircle is a right angle.

- Angles in the same segment are equal.

- The sum of the opposite angles of a cyclic quadrilateral is 180°.
  (A quadrilateral whose vertices (corners) all lie on the circumference of a circle is called a cyclic quadrilateral.)

  $a + c = 180°$ and $b + d = 180°$

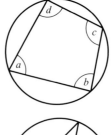

- The angle between the chord and the tangent at the point of contact is equal to the angle in the alternate segment. (Alternate segment theorem)

P, Q, R and S are points on the circumference of a circle, centre O.

PR is a diameter of the circle.
Angle PSQ = 56°.

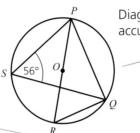

Diagram **NOT** accurately drawn

a Find the size of angle PQR.
Give a reason for your answer.
b Find the size of angle PRQ.
Give a reason for your answer.
c Find the size of angle POQ.
Give a reason for your answer.

(1387 November 2003)

a Angle PQR = 90°
(The angle in a semicircle is a right angle)

b Angle PRQ = 56°
(Angles in the same segment)

c Angle POQ = 2 × 56° = 112°
(The angle at the centre is twice the angle at the circumference)

EXAM TIP

Diagram **NOT** accurately drawn means that taking measurements from the diagram will give WRONG answers.

'Find' means that it may or may not involve some working. In fact, parts **a** and **b** are 'Write down' and part **c** is 'Calculate' or 'Work out'.

EXAM TIP

Reasons must contain all the key points expressed in mathematical language, but do not have to be written as complete sentences.

P, Q, R and S are points on the circumference of a circle.
PT is a tangent to the circle at P.
Angle SPT = 43°.
Angle PQR = 82°.

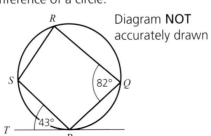

Diagram **NOT** accurately drawn

a Find the size of angle PRS.
Give a reason for your answer.
b Find the size of angle PSR.
Give a reason for your answer.

a Angle PRS = 43°
(Alternate segment theorem)

The reason may be shortened to this.

b Angle PSR = 180° − 82° = 98°
(The sum of the opposite angles of a cyclic quadrilateral is 180°)

1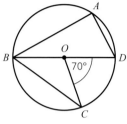

Diagram **NOT** accurately drawn

A, B, C and D are points on the circumference of a circle, centre O.
BOD is a straight line.

a   Find the size of angle BAD. Give a reason for your answer.
b   Find the size of angle CBD. Give a reason for your answer.

(1387 June 2006)

2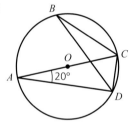

Diagram **NOT** accurately drawn

A, B, C and D are points on the circumference of a circle, centre O.
AC is a diameter of the circle.
Angle $DAC = 20°$.

a   Find the size of angle ACD.
b   Find the size of angle DBC. Give a reason for your answer.

(1387 November 2006)

3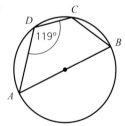

Diagram **NOT** accurately drawn

A, B, C and D are points on the circumference of a circle.
AB is a diameter of the circle.
Angle $ADC = 119°$.

a   i   Work out the size of angle ABC.
    ii   Give a reason for your answer.
b   Work out the size of angle BAC.

(4400 November 2006)

4

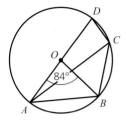

Diagram **NOT**
accurately drawn

A, *B*, *C* and *D* are points on the circumference of a circle with centre *O*.
AOD is a diameter of the circle. Angle *AOB* = 84°.

a i Calculate the size of angle *ACB*.
   ii Give a reason for your answer.
b Calculate the size of angle *BCD*.

(4400 November 2005)

5

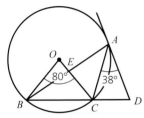

Diagram **NOT**
accurately drawn

Points *A*, *B*, *C* and *D* lie on the circumference of a circle with centre *O*.
DA is a tangent to the circle at *A*. *BCD* is a straight line.
OC and *AB* intersect at *E*. Angle *BOC* = 80°. Angle *CAD* = 38°.

a Calculate the size of angle *BAC*.
b Calculate the size of angle *OBA*.
c Give a reason why it is not possible to draw a circle with diameter *ED*
   through *A*.

(1386 November 2000)

6

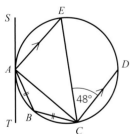

Diagram **NOT**
accurately drawn

A, *B*, *C*, *D* and *E* are points on the circumference of a circle.
SAT is a tangent to the circle at *A*.
AB = *BC*. *AE* is parallel to *CD*. Angle *ECD* = 48°.

a Calculate the size of angle *ABC*. Give reasons for your answer.
b Calculate the size of angle *TAB*. Give reasons for your answer.

(1386 November 1999)

> **HINT**
>
> 'Reasons' means that more
> than one answer is required
> in each part. Some reasons
> require general angle facts.

SKILL

Use vectors to solve geometric problems in two dimensions

EXAM FACTS

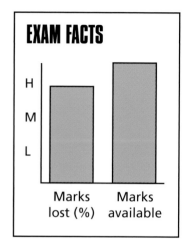

KEY FACTS

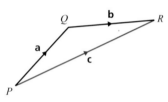

In vector work, numbers are called scalars.

- Triangle law of vector addition
  $\vec{PQ} + \vec{QR} = \vec{PR}$ or a + b = c

- When $\vec{PQ}$ = a, $\vec{QP}$ = −a

- When $\vec{PQ} = k\vec{RS}$, where k is a positive scalar, the lines PQ and RS are parallel and the length of PQ is k times the length of RS.

- When $\vec{PQ} = k\vec{PR}$, the lines PQ and PR are parallel. But the point P is common to both lines so that PQ and PR are part of the same straight line. That is, P, Q and R lie on the same straight line.

- 'In the same direction' and 'parallel' are not the same. Two vectors can be parallel but in opposite directions.

  So if $\vec{PQ} = 4\vec{RS}$, then $\vec{PQ}$ and $\vec{RS}$ are in the same direction.

  But if $\vec{PQ} = -4\vec{RS}$, then $\vec{PQ}$ and $\vec{RS}$ are in opposite directions.

- The vector a can be written as a column vector $\begin{pmatrix} 3 \\ 2 \end{pmatrix}$.

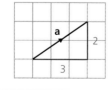

OCD is a triangle.
A is a point on OC
B is a point on OD

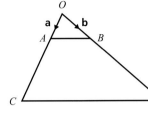

Diagram **NOT** accurately drawn

$\overrightarrow{OA} = $ a
$\overrightarrow{OB} = $ b
$\overrightarrow{OC} = 4$a
$\overrightarrow{BD} = 3$b

a Express in terms of **a** and **b**
   i $\overrightarrow{AB}$ ii $\overrightarrow{CD}$

b Hence show that CD is parallel to AB.

c Write down the value of $\frac{CD}{AB}$.

Express $\overrightarrow{AB}$ and $\overrightarrow{CD}$ in terms of known vectors using the triangle law.

a i $\overrightarrow{AB} = \overrightarrow{AO} + \overrightarrow{OB}$

   $\overrightarrow{AO} = \underline{a}$ as $\overrightarrow{OA} = \underline{a}$

   $\overrightarrow{AB} = -\underline{a} + \underline{b} = \underline{b} - \underline{a}$

ii $\overrightarrow{CD} = \overrightarrow{CO} + \overrightarrow{OD}$

   $\overrightarrow{CO} = -4\underline{a}$ as $\overrightarrow{OC} = 4\underline{a}$

   $\overrightarrow{OD} = \overrightarrow{OB} + \overrightarrow{BD} = \underline{b} + 3\underline{b} = 4\underline{b}$

   $\overrightarrow{CD} = -4\underline{a} + 4\underline{b} = 4\underline{b} - 4\underline{a}$

b $\overrightarrow{CD} = 4\underline{b} - 4\underline{a} = 4(\underline{b} - \underline{a}) = 4\overrightarrow{AB}$

   Hence CD is parallel to AB

c $CD = 4AB$ so $\frac{CD}{AB} = 4$

In print vectors are shown in heavy type, e.g. **a**, which is not easy to write.
The correct notation in written work is $\underline{a}$.

To show that CD is parallel to AB, show that $\overrightarrow{CD} = k\overrightarrow{AB}$, for a number k.

WARNING

Vectors cannot be divided so, to write down the value of $\frac{CD}{AB}$, you need to use the lengths (magnitudes) of the vectors.

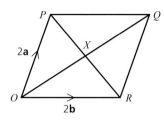

Diagram **NOT** accurately drawn

$OPQR$ is a parallelogram with PQ parallel to OR.

$\overrightarrow{OP} = 2\mathbf{a}$ $\overrightarrow{OR} = 2\mathbf{b}$

X is the midpoint of PR.

a Find the vector $\overrightarrow{PX}$ in terms of $\mathbf{a}$ and $\mathbf{b}$.
b Prove that X is the midpoint of OQ.

(5540 June 2005)

a   As X is the midpoint of PR, $\overrightarrow{PX} = \frac{1}{2}\overrightarrow{PR}$

Express $\overrightarrow{PR}$ in terms of known vectors using the triangle law.

$\overrightarrow{PR} = \overrightarrow{PO} + \overrightarrow{OR}$

$\overrightarrow{PR} = -2\underline{a} + 2\underline{b} = 2\underline{b} - 2\underline{a} = 2(\underline{b} - \underline{a})$

So $\overrightarrow{PX} = \frac{1}{2} \times 2(\underline{b} - \underline{a}) = \underline{b} - \underline{a}$

To prove that X is the midpoint of OQ, write both $\overrightarrow{OX}$ and $\overrightarrow{OQ}$ in terms of **a** and **b**.

b $\overrightarrow{OX} = \overrightarrow{OP} + \overrightarrow{PX} = 2\underline{a} + \underline{b} - \underline{a} = \underline{a} + \underline{b}$

$\overrightarrow{OQ} = \overrightarrow{OP} + \overrightarrow{PQ} = \overrightarrow{OP} + \overrightarrow{OR} = 2\underline{a} + 2\underline{b} = 2(\underline{a} + \underline{b})$

So $\overrightarrow{OX} = \frac{1}{2}\overrightarrow{OQ}$

$\overrightarrow{PQ} = \overrightarrow{OR}$ as $OPQR$ is a parallelogram, meaning that its opposite sides are equal and parallel.

This means that the lines OX and OQ are parallel but the point O is common to both lines so that O, X and Q lie on the same line with $OX = \frac{1}{2}OQ$.

Hence X is the midpoint of OQ.

1 $ABCD$ is a parallelogram
AB is parallel to DC. AD is parallel to BC
$\overrightarrow{AB} = \mathbf{p}$ $\overrightarrow{AD} = \mathbf{q}$

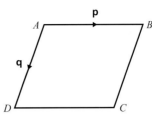

Diagram **NOT** accurately drawn

 a   Express, in terms of $\mathbf{p}$ and $\mathbf{q}$,
   i $\overrightarrow{AC}$ ii $\overrightarrow{BD}$

AC and BD are diagonals of parallelogram $ABCD$.
AC and BD intersect at T.

 b   Express $\overrightarrow{AT}$ in terms of $\mathbf{p}$ and $\mathbf{q}$.

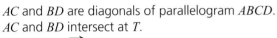

Diagram **NOT** accurately drawn

(1387 June 2006)

2 OAB is a triangle.
$\overrightarrow{OA} = \mathbf{a}$ $\overrightarrow{OB} = \mathbf{b}$
P is the point on AB such that $AP : PB = 2 : 1$
Write $\overrightarrow{OP}$ in terms of $\mathbf{a}$ and $\mathbf{b}$.

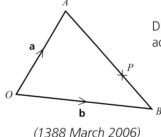

Diagram **NOT** accurately drawn

(1388 March 2006)

3 $PQRS$ is a parallelogram.
X is the midpoint of QR and Y is the midpoint of SR.
$\overrightarrow{PQ} = \mathbf{a}$ and $\overrightarrow{PS} = \mathbf{b}$.

Diagram **NOT** accurately drawn

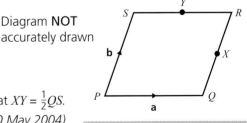

 a   Write down, in terms of $\mathbf{a}$ and $\mathbf{b}$, expressions for
   i $\overrightarrow{PX}$,   ii $\overrightarrow{PY}$,   iii $\overrightarrow{QS}$

 b   Use a vector method to show that XY is parallel to QS and that $XY = \frac{1}{2}QS$.
(4400 May 2004)

4 PQR is a triangle. M and N are the midpoints of PQ and PR respectively.
$\overrightarrow{PM} = \mathbf{a}$ $\overrightarrow{PN} = \mathbf{b}$
 a   Find in terms of $\mathbf{a}$ and/or $\mathbf{b}$,
   i $\overrightarrow{MN}$,   ii $\overrightarrow{PQ}$,   iii $\overrightarrow{QR}$

 b   Use your answers to
 a i and **iii** to write down
 two geometrical facts about
 the lines MN and QR.

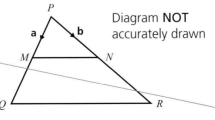

Diagram **NOT** accurately drawn

(4400 May 2005)

EXAM TIP

The wording 'in terms of **a** and/or **b**' can be used on examination papers when an answer may or may not involve both **a** and **b**, as in part **a** here, even though both **a** and **b** do not appear in each answer.

5 *OAB* is a triangle. *B* is the midpoint of *OR*. *Q* is the midpoint of *AB*.
$\overrightarrow{OP} = 2\mathbf{a}$ $\overrightarrow{PA} = \mathbf{a}$ $\overrightarrow{OB} = \mathbf{b}$

  a Find, in terms of **a** and **b**, the vectors
    i $\overrightarrow{AB}$,   **ii** $\overrightarrow{PR}$,   **iii** $\overrightarrow{PQ}$.

  b Hence explain why *PQR* is a straight line.

  The length of *PQ* is 3 cm.
  c Find the length of *PR*.

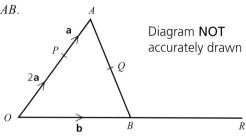

Diagram **NOT** accurately drawn

(1387 November 2006)

6 The diagram shows two triangles *OAB* and *OCD*.
OAC and *OBD* are straight lines.
AB is parallel to *CD*.
$\overrightarrow{OA} = \mathbf{a}$ and $\overrightarrow{OB} = \mathbf{b}$.

The point *A* cuts the line *OC* in the ratio *OA* : *OC* = 2 : 3.
Express $\overrightarrow{CD}$ in terms of **a** and **b**.

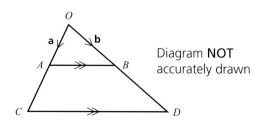

Diagram **NOT** accurately drawn

(1385 November 2001)

7 *ABCD* is a quadrilateral.
K is the midpoint of *AB*.
L is the midpoint of *BC*.
M is the midpoint of *CD*.
N is the midpoint of *AD*.
$\overrightarrow{AK} = \mathbf{a}$, $\overrightarrow{AN} = \mathbf{b}$ and $\overrightarrow{DM} = \mathbf{c}$.

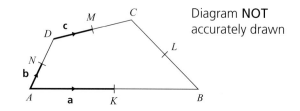

Diagram **NOT** accurately drawn

  a Find, in terms of **a**, **b** and **c**, the vectors
    i $\overrightarrow{KN}$,   **ii** $\overrightarrow{AC}$,   **iii** $\overrightarrow{BC}$,   **iv** $\overrightarrow{LM}$

  b Write down two geometrical facts about the lines *KN* and *LM*
      which could be deduced from your answers to part **a**.

(1385 June 2002)

8 *PQRS* is a trapezium in which $\overrightarrow{PQ} = \mathbf{a}$, $\overrightarrow{SR} = 2\mathbf{a}$ and $\overrightarrow{SP} = \mathbf{b}$.
  a Express in terms of **a** and **b**,
    i $\overrightarrow{SQ}$ **ii** $\overrightarrow{RQ}$

  The point *T* on *SQ* is such that $\overrightarrow{ST} = 2\overrightarrow{TQ}$.
  b Express in terms of **a** and **b**,
    i $\overrightarrow{QT}$ **ii** $\overrightarrow{PT}$ **iii** $\overrightarrow{TR}$

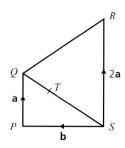

Diagram **NOT** accurately drawn

  c Use your results for **b ii** and **b iii**
    i   to explain why the points *P*, *T* and *R* lie on a straight line,
    ii to find the value of $\dfrac{PT}{PR}$.

9 *PQRS* is a parallelogram in which $\overrightarrow{PQ} = \mathbf{a}$ and $\overrightarrow{PS} = \mathbf{b}$.
   PQT is a straight line where Q is the midpoint of *PT*.

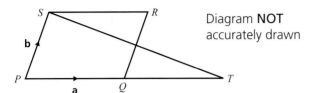

Diagram **NOT** accurately drawn

   a   Express in terms of **a** and **b**,
      i $\overrightarrow{PR}$,   ii $\overrightarrow{ST}$.

X is the point on *PR* such that $\overrightarrow{PX} = \frac{2}{3}\overrightarrow{PR}$

   b   Express the vector $\overrightarrow{SX}$ in terms of **a** and **b**.
   c   Hence show that X lies on *ST* and state the ratio in which X divides *ST*.

10 *ABCD* is a trapezium in which *AB* and *DC* are parallel.
   $\overrightarrow{AB} = \begin{pmatrix} 3 \\ 5 \end{pmatrix}$ and $\overrightarrow{DC} = \begin{pmatrix} 6 \\ k \end{pmatrix}$

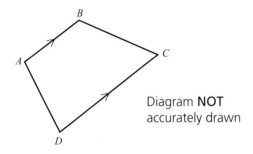

Diagram **NOT** accurately drawn

   a   Find the value of the scalar k.

$\overrightarrow{BC} = \begin{pmatrix} 6 \\ -5 \end{pmatrix}$

   b   Find, as a column vector, i $\overrightarrow{AC}$,   ii $\overrightarrow{AD}$.

The point T lies on *AC* so that $\overrightarrow{AT} = \begin{pmatrix} 3 \\ 0 \end{pmatrix}$.

   c   Show that T lies on *BD*.

SKILLS

Draw a histogram from given information
Use a histogram to complete a frequency table

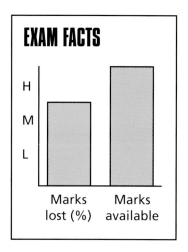

KEY FACTS

- In a histogram, the area of a bar gives the frequency of the class interval.
- To draw a histogram from a table
  - work out the width of each class interval (the class width)
  - work out the frequency density, which gives the height of each bar, by using

    $$\text{frequency density} = \frac{\text{frequency}}{\text{class width}}$$

- To complete a table from a histogram
  - work out the area of each bar, this represents the frequency of the class interval

    frequency = frequency density × class width

Getting it right

Kath recorded the times, in minutes, taken by 170 students to travel to school.

The table gives information about her results.

Time (t minutes)	Frequency
$0 \leqslant t < 20$	70
$20 \leqslant t < 35$	45
$35 \leqslant t < 45$	44
$45 \leqslant t < 50$	11

Use the information in the table to draw a histogram.

(1388 November 2005)

 WARNING

The most common error is to use the frequency for the heights of the bars. As the class intervals are not all the same, the frequency density must be calculated and used for the heights of the bars.

Time (t minutes)	Frequency	Class width	$\dfrac{\text{Frequency}}{\text{Class width}}$
$0 \leqslant t < 20$	70	$20 - 0 = 20$	$\dfrac{70}{20} = 3.5$
$20 \leqslant t < 35$	45	$35 - 20 = 15$	$\dfrac{45}{15} = 3$
$35 \leqslant t < 45$	44	$45 - 35 = 10$	$\dfrac{44}{10} = 4.4$
$45 \leqslant t < 50$	11	$50 - 45 = 5$	$\dfrac{11}{5} = 2.2$

Work out each class width.

Work out the frequency density.

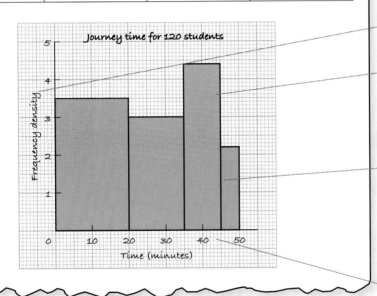

Journey time for 120 students

Label the vertical axis 'frequency density'.

Draw the histogram. Use the frequency density for the heights of the bars.

WARNING

A common error is to leave gaps between the bars. The data is continuous so there should not be any gaps between bars.

Look carefully at the frequency table for the width of the bars.

The histogram and table show information about the number of emails received by each of the students in a school.

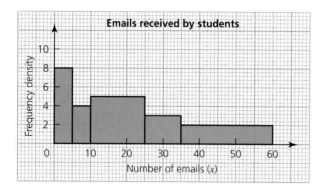

Emails received by students

Number of emails (x)	Frequency
$0 < x \leqslant 5$	
$5 < x \leqslant 10$	20
$10 < x \leqslant 25$	
$25 < x \leqslant 35$	
$35 < x \leqslant 60$	

Use the information in the histogram to complete the table.

(1388 March 2006)

WARNING

The most common error is to read the frequency density values from the vertical axis and give these as the frequency.

Work out the class width.

Read the frequency density from the vertical axis on the graph.

Use frequency = class width × frequency density to work out the frequency.

Method 1

Number of emails (x)	Frequency	Class width	Frequency density
$0 < x \leqslant 5$	5 × 8 = 40	5	8
$5 < x \leqslant 10$	20		
$10 < x \leqslant 25$	15 × 5 = 75	15	5
$25 < x \leqslant 35$	10 × 3 = 30	10	3
$35 < x \leqslant 60$	25 × 2 = 50	25	2

Method 2

The frequency for the class interval $5 < x \leqslant 10 = 20$

Area of this bar $= 2\,cm^2$ (if drawn full size)

1 cm² represents $20 \div 2 = 10$ emails

Number of emails (x)	Frequency	Area of bar (cm²)
$0 < x \leqslant 5$	$4 \times 10 = 40$	4
$5 < x \leqslant 10$	20	
$10 < x \leqslant 25$	$7.5 \times 10 = 75$	7.5
$25 < x \leqslant 35$	$3 \times 10 = 30$	3
$35 < x \leqslant 60$	$5 \times 10 = 50$	5

Work out the area of the bar whose frequency is given.

Divide the frequency by the area to work out the number of emails 1 cm² represents.

Multiply each area by 10 to work out the frequency represented by each bar.

Now try these

1 The table gives information about the heights, in centimetres, of some 15 year old students.

Height (h cm)	$145 < h \leqslant 155$	$155 < h \leqslant 175$	$175 < h \leqslant 190$
Frequency	10	80	24

Use the table to draw a histogram.

(1388 January 2003)

2 The table shows the distribution of the ages of passengers travelling on a plane from London to Belfast.

Age (x years)	Frequency
$0 < x \leqslant 20$	28
$20 < x \leqslant 35$	36
$35 < x \leqslant 45$	20
$45 < x \leqslant 65$	30

Draw a histogram to show this distribution.

(1388 March 2003)

3 The histogram gives information about the weights, in kilograms,
   of some boxes.

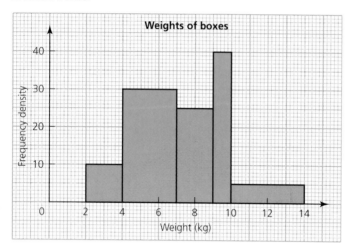

Use the histogram to complete the table.

Weight (w kg)	Frequency
$2 \leqslant w < 4$	20
$4 \leqslant w < 7$	
$7 \leqslant w < 9$	
$9 \leqslant w < 10$	
$10 \leqslant w < 14$	

4

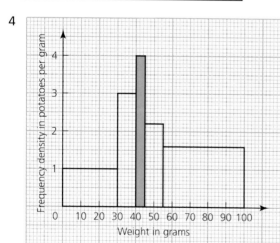

The histogram gives information about the weights of some potatoes.
The shaded bar represents 20 potatoes.
a   Work out how many of the potatoes weigh 30 grams or less.
b   Work out how many of the potatoes weigh more than 45 grams.

(1388 January 2005)

5 The table gives information about the number of hours worked by some factory workers.

Number of hours (n) worked	Frequency
$0 < n \leqslant 5$	15
$5 < n \leqslant 15$	42
$15 < n \leqslant 35$	40
$35 < n \leqslant 50$	6

Use the table to draw a histogram.

(1388 March 2006)

6 The table gives information about the times, in minutes, some students took to travel to school one day.

Time (t) in minutes	Frequency
$0 < t \leqslant 25$	50
$25 < t \leqslant 35$	40
$35 < t \leqslant 50$	15

Use this information to draw a histogram.

(1388 November 2006)

7 The table shows some information about the lengths, in cm, of 60 babies.

Length (x cm)	Frequency
$20 < x \leqslant 30$	10
$30 < x \leqslant 45$	30
$45 < x \leqslant 50$	20

Draw a histogram for this information.

(1388 March 2007)

SKILL

Use stratified sampling in problems

EXAM FACTS

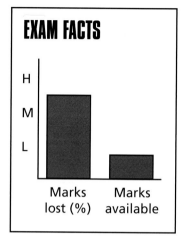

KEY FACTS

When a sample is to be taken from a population, it might be possible to split the population into a number of groups, called strata (singular stratum). For example, strata might be age, gender or year group in a school.

- In a stratified sample, the fraction of the size of each stratum in the sample to the size of the sample is the same as the fraction of the size of the stratum in the population to the size of the population.

The members of each stratum in the sample are chosen at random from the members of the stratum in the population. A random sample is one in which each member of the population being sampled is equally likely to be selected.

- When the size of a stratum in the poulation is X and the size of the population is N, for a sample of size n, the size of the stratum in the sample is s where

$$\frac{s}{n} = \frac{X}{N} \text{ or } s = \frac{X}{N} \times n$$

- A stratified sample ensures that the sample is representative of the population.

Getting it right

	Year 10 Group	Year 11 Group
Boys	100	50
Girls	90	60

The table shows the number of boys and the number of girls in Year 10 and Year 11 of a school.
A stratified sample of size 50 is to be taken from Year 10 and Year 11.

Calculate the number of girls to be sampled from Year 11.

$100 + 50 + 90 + 60 = 300$

Number of students in Years 10 and 11 $= 300$

Fraction of population that are Year 11 girls $= \frac{60}{300} = \frac{1}{5}$

Number of Year 11 girls in the sample $= \frac{1}{5} \times 50 = 10$

This is the size of the population.

This is the fraction of the population that are Year 11 girls. It is also the fraction of the sample that should be Year 11 girls.

50 is the size of the sample.

The table shows some information about the members of a golf club.

Age range	Male	Female	Total
Under 18	29	10	39
18 to 30	82	21	103
31 to 50	147	45	192
Over 50	91	29	120
	Total number of members		454

The club secretary carries out a survey of the members. He chooses a sample, stratified both by age range and by gender, of 90 of the 454 members.

Work out an estimate of the number of members that are male in the age range 31 to 50, he would have to sample.

(1388 June 2003)

Number of males aged 31 to 50 in the population $= 147$

Fraction of the population that are males in the age range 31 to

$50 = \frac{147}{454}$

Number of males required for this stratum

$= \frac{147}{454} \times 90 = 29.14...$

Number of males in the sample in the age range 31 to 50 $= 29$

EXAM TIP

The word "estimate" is used here because answers sometimes require rounding. It does not mean "have a guess".

90 is the size of the sample.

WARNING

The number of males aged 31 to 50 must be a whole number. A common error is to give unrounded answers. Rounding to the nearest whole number gives the answer 29 but 30 is also acceptable.

1 There are 1000 students in Nigel and Sonia's school.
This table shows the gender and the number of students in each year group.

Year group	Number of boys	Number of girls	Total
7	100	100	200
8	90	80	170
9	120	110	230
10	80	120	200
11	100	100	120

Sonia is carrying out a survey about how much homework students are given.

She decides to take a stratified sample of 100 students from the whole school.

Calculate how many in the stratified sample should be
i students from Year 9,
ii boys from Year 10.

(1385 June 2000)

2 The table shows the number of students in each year group at a school.

Year Group	Number of students
7	270
8	270
9	240
10	200
11	180

Mrs Fox is the music teacher. She is carrying out a survey about the students' favourite radio channel.
She uses a stratified sample of 40 students according to year group.
Calculate how many Year 10 students should be in her sample.

(1385 November 2002)

3 The table gives information about the number of girls in each of four schools.

School	A	B	C	D	Total
Number of girls	126	82	201	52	461

Jenny did a survey of these girls. She used a stratified sample of exactly 80 girls according to school.
Work out the number of girls from each school that were in her sample of 80.
Complete the table.

School	A	B	C	D	Total
Number of girls					80

(1388 June 2006)

4 The table shows the number of students in three groups attending Maths City High School last Monday. No student belonged to more than one group.
Mrs Allen interviewed some of these students. She used a stratified sample of 50 students according to each group.
Work out the number of students from each group which should have been in her sample of 50.

Group	Number of students
A	135
B	225
C	200

(1385 Summer 2001)

5

	Year 10 Group	Year 11 Group
Boys	75	50
Girls	65	25

The table shows the number of boys and the number of girls in Year 10 and in Year 11 of a school.
The headteacher of the school wants to find out what the pupils think about their new healthy meals.

A sample of size 50 is taken, stratified by both year group and by gender.
a   Calculate the number of Year 11 girls in the sample.

Trevor says that the number of Year 10 boys in the sample is three times the number of Year 11 girls in the sample.
b   Is Trevor correct? You must show how you reached your decision.

6 Mrs Green wants to find out how often people visit her sports centre. The sports centre has 5000 members. Their ages are from 10 years to 60 years. The table shows some information about these members.

Age (years)	Number of males	Number of females
10 to 16	1500	1300
17 to 25	600	400
26 to 40	750	200
41 to 60	150	100

Mrs Green takes a sample of 200 of the 5000 members. Her sample is stratified by both age and gender.

Calculate the number of males aged from 26 years to 40 years in her sample.

(2544 March 2007)

7 A school has 450 students. Each student studies one of Greek or Spanish or German or French. The table shows the number of students who study each of these languages.

An inspector wants to look at the work of a stratified sample of 70 of these students. Find the number of students studying each of these languages that should be in the sample.

Language	Number of students
Greek	45
Spanish	121
German	98
French	186

(1387 November 2006)

8 Mathstown College has 500 students, all of them in the age range 16 to 19. The incomplete table shows information about the students.

Age (years)	Number of male students	Number of female students
16	50	30
17	60	40
18	76	54
19		

A newspaper reporter is carrying out a survey into students' part-time jobs. She takes a sample, stratified both by age and by gender, of 50 of the 500 students.

a Calculate the number of 18 year old male students to be sampled.

In the sample, there are 9 female students whose age is 19 years.

b Work out the least number of 19 year old female students in the college.

(1387 November 2003)

9 The table shows some information about the pupils at Statson School.

Year group	Boys	Girls	Total
Year 7	104	71	175
Year 8	94	98	192
Year 9	80	120	200
Total	278	289	567

Kelly carries out a survey of the pupils at Statson School.
She takes a sample of 80 pupils, stratified by both Year group and gender.
a   Work out the number of Year 8 boys in her sample.
b   i   Explain what is meant by a random sample.
    ii Describe a method that Kelly could use to take a random sample of
        Year 8 boys.

(1388 June 2005)

10 a   Explain what is meant by a stratified sample.
   b   The table shows the number of students in each year group at
       Mathstown High School.

Year Group	Number of students
9	180
10	160
11	175
12	250
13	230

Karl is carrying out a survey about the students' favourite television
programmes.
He uses a stratified sample of 50 students according to year group.

Calculate the number of Year 13 students which should be in his sample.

28 Using P(A and B) = P(A) × P(B)

SKILL

Use multiplication of probabilities for two independent outcomes

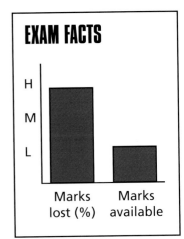
KEY FACTS

- When the outcomes, A and B, of two events are independent

  P(A and B) = P(A) × P(B)

- A probability tree diagram shows all the possible outcomes of more than one event by following all the possible paths along the branches of the tree. When moving along each path, multiply the probabilities on each of the branches.

Getting it right

A box contains 5 wood drills and 3 metal drills only.

Sue takes a drill at random from the box and then replaces it. Cliff then takes a drill at random from the box.

Work out the probability that Sue and Cliff will each take a wood drill.

P(Sue takes a wood drill) = $\frac{5}{8}$

P(Cliff takes a wood drill) = $\frac{5}{8}$

P(Sue takes a wood drill **and** Cliff takes a wood drill)

$= \frac{5}{8} \times \frac{5}{8}$

$= \frac{25}{64}$

> Since the events are independent use P(A and B) = P(A) × P(B). This gets 1 mark.

WARNING

A common error is to add the probabilities instead of multiplying. This would give a probability greater than 1 which is impossible.

In a café, the probability that a customer orders chips is 0.7
In the same café, the probability that a customer orders coffee is 0.2
The two events are independent.
This information is shown on the probability tree diagram

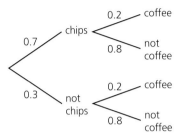

Work out the probability that a customer will order both chips and coffee.

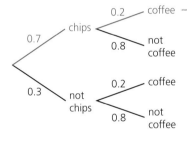

Identify the path to get chips and coffee.

$P(\text{chips and coffee}) = P(\text{chips}) \times P(\text{coffee})$
$= 0.7 \times 0.2$
$= 0.14$

When moving along a path multiply the probabilities on each of the branches to give 0.7×0.2
This gets 1 mark.

Now try these

1 Nick and Dravid each try to convert a rugby penalty kick.
The probability that Nick will convert his penalty kick is 0.75
The probability that Dravid will convert his penalty kick is 0.9
Work out the probability that Nick and Dravid will both convert their penalty kicks.

2 Here is a biased spinner.
When the pointer is spun, the score is 1 or 2 or 3 or 4
The probability that the score is 1 is 0.3
The probability that the score is 2 is 0.6
Nassim spins the pointer twice.
Work out the probability that the score is 1 both times.

(4400 May 2006)

3 Simon plays one game of tennis and one game of snooker.
The probability that Simon will win at tennis is $\frac{3}{4}$
The probability that Simon will win at snooker is $\frac{1}{3}$

   a Complete the probability tree diagram.

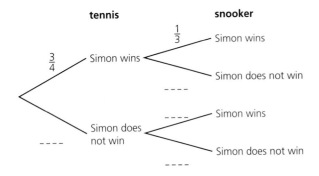

   b Work out the probability that Simon wins both games.

(1387 June 2006)

4 The probability that a washing machine will break down in the first 5 years
of use is 0.27
The probability that a television will break down in the first 5 years of
use is 0.17
Mr Khan buys a washing machine and a television on the same day.
By using a tree diagram or otherwise, calculate the probability that, in
five years after that day, both the washing machine and the television
will break down.

(1385 June 1998)

5 The probability of a car chosen at random having

   defective tyres is 0.065
   defective brakes is 0.04

   Work out the probability that a car chosen at random
   i will have defective tyres and defective brakes,
   ii will not have defective tyres and will not have defective brakes.

6 The diagram shows six counters.

   Each counter has a letter on it.
   Bishen puts the six counters into a bag.
   He takes a counter at random from the bag.
   He records the letter which is on the counter and replaces the counter in
   the bag.
   He then takes a second counter at random and records the letter which is
   on the counter.

   Calculate the probability that the first letter will be A and the second letter
   will be N.

(4400 May 2005)

29 Using P(A or B) = P(A) + P(B)

SKILL

Use addition of probabilities for two mutually exclusive outcomes

EXAM FACTS

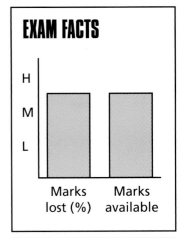

For a reminder of how to calculate P(A and B) turn to pages 116 and 117.

KEY FACTS

- When the outcomes, A and B, of an event are mutually exclusive

  P(A or B) = P(A) + P(B)

- A probability tree diagram shows all of the possible outcomes of more than one event by following all of the possible paths along the branches of the tree. When two paths are mutually exclusive, add the probabilities that have been calculated by multiplying the probabilities along each path.

Getting it right

Jan's bus can be on time or early or late. The probability that her bus will be early is 0.18 The probability that her bus will be on time is 0.4
Work out the probability that Jan's bus will be either early or on time.

P(Jan's bus will be either early or on time.)

= P(Jan's bus will be early) + P(Jan's bus will be on time)

= 0.18 + 0.4

= 0.58

The outcomes are mutually exclusive (they cannot happen at the same time) so add the probabilities. This gets 1 mark.

WARNING

A common error is to make an arithmetic mistake when adding 0.18 and 0.4
For example 0.22 is a common answer.

Jo plays one game of tennis and one game of squash. The probability that Jo will win at tennis is 0.6. The probability that Jo will win at squash is 0.75 The probability tree diagram shows the possible outcomes.

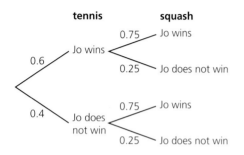

Work out the probability that Jo wins only one game.

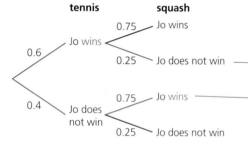

Identify the paths where Jo wins only one of the games.

Jo wins only the tennis game.

Jo wins only the squash game.

P(Jo wins at tennis and does not win at squash)
= 0.6 × 0.25
= 0.15

P(Jo does not win at tennis and wins at squash)
= 0.4 × 0.75
= 0.3

When moving along a path multiply the probabilities on each of the branches to give
0.6 × 0.25
and 0.4 × 0.75
Either of these gets 1 mark.

P(Jo wins only one game)
  = P(Jo wins at tennis and does not win at squash)
    +
    P(Jo does not win at tennis and wins at squash)
  = 0.15 + 0.3
  = 0.45

The paths represent outcomes which are mutually exclusive so add the probabilities. This gets 1 mark.

A box contains 5 wood drills and 3 metal drills only.
Sue takes a drill at random from the box and then replaces it.
Cliff then takes a drill at random from the box.
Work out the probability that at least one metal drill will be taken

Method 1

P(Sue takes a metal drill and Cliff takes a wood drill)

$$= \frac{3}{8} \times \frac{5}{8} = \frac{15}{64}$$

P(Cliff takes a metal drill and Sue takes a wood drill)

$$= \frac{3}{8} \times \frac{5}{8} = \frac{15}{64}$$

P(both take a metal drill) $= \frac{3}{8} \times \frac{3}{8} = \frac{9}{64}$

P(at least one metal drill will be taken)

$$= \frac{15}{64} + \frac{15}{64} + \frac{9}{64}$$

$$= \frac{39}{64}$$

Method 2

P(at least one metal drill will be taken)

$$= 1 - P(both\ take\ a\ wood\ drill)$$

$$= 1 - \frac{5}{8} \times \frac{5}{8}$$

$$= 1 - \frac{25}{64}$$

$$= \frac{39}{64}$$

For at least one metal drill to be taken either Sue takes a metal drill and Cliff takes a wood drill OR Cliff takes a metal drill and Sue takes a wood drill OR they both take a metal drill. The 3 events are mutually exclusive.

Find the probability that each of the 3 events occur using P(A and B) = P(A) × P(B). This gets 1 mark when correctly used for the first time.

Since the 3 outcomes are mutually exclusive use P(A or B) = P(A) + P(B). This gets 1 mark.

For at least one metal drill to be taken, they cannot both take a wood drill.

Find P(both take a wood drill) using P(A and B) = P(A) × P(B). Then use 1 − p to find the probability of this NOT happening.

EXAM TIP

You will not be penalised if you do this the long way, (metal and metal + metal and wood + wood and metal) but you are more likely to make a mistake.

1 Nick and Dravid each try to convert a rugby penalty kick.
The probability that Nick will convert his penalty kick is 0.75
The probability that Dravid will convert his penalty kick is 0.9
Work out the probability that just one of the penalty kicks will be
converted.

2 A factory produces electric switches.
The probability that a switch is faulty is 0.15
Two switches are chosen at random.
   a Copy and complete the probability tree diagram to show the possible
      outcomes.

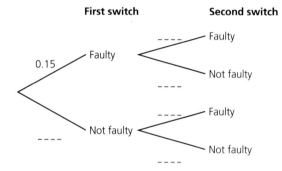

   b Calculate the probability that at least one switch is faulty.

3 The diagram shows six counters.

Each counter has a letter on it.

Bishen puts the six counters into a bag.
He takes a counter at random from the bag.
He records the letter which is on the counter and replaces the counter in
the bag.
He then takes a second counter at random and records the letter which is
on the counter.

Calculate the probability that both letters will be the same

(4400 May 2005)

4 The probability that a washing machine will break down in the first 5 years of use is 0.27
The probability that a television will break down in the first 5 years of use is 0.17
Mr Khan buys a washing machine and a television on the same day.
By using a tree diagram or otherwise, calculate the probability that, in five years after that day, at least one of them will break down.

(1385 June 1998)

5 Here is a biased spinner.

When the pointer is spun, the score is 1 or 2 or 3 or 4
The probability that the score is 1 is 0.3
The probability that the score is 2 is 0.6

a Work out the probability that the score is 1 or 2
b Work out the probability that the score is 3 or 4

Nassim spins the pointer twice.
c Work out the probability that the score is 2 exactly once.

(4400 May 2006)

SKILL

Solve conditional probability problems

EXAM FACTS

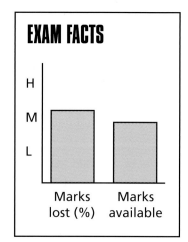

KEY FACT

- To solve conditional probability problems remember that the outcome of an event is dependent on the outcome of a previous event.
  For example, choosing two pieces of fruit **without replacing** the first one where the choice of the second piece of fruit is dependent on the choice of the first.

REFERENCE

For a reminder of how to deal with independent outcomes turn to pages 116 and 117.

For a reminder of how to deal with mutually exclusive outcomes to pages 119 to 121.

Getting it right

On journeys between home and school Jaspal either walks or catches a bus.

The probability that Jaspal will walk to school is 0.6

If Jaspal walks to school, the probability that she will walk back home is 0.85

If she catches a bus to school, the probability that she will walk back home is 0.45

a   Calculate the probability that, on a particular day, Jaspal will walk to school and will also walk back home.

b   Calculate the probability that, on a particular day, Jaspal will catch a bus back home from school.

(1385 November 1999)

'If' indicates conditional probability

WARNING

Those who do not draw a probability tree diagram often fail to realise that this is a conditional probability question.

a

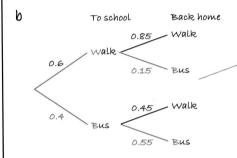

To school   Back home

Walk   0.85   Walk
0.6   0.15   Bus

Bus   0.45   Walk
0.4   0.55   Bus

$P(WW) = 0.6 \times 0.85 = 0.51$

It is best to draw a probability tree diagram in this case. The diagram is completed by using P(Bus) = P(NOT Walk) = 1 − P(Walk) to find 0.4, 0.15 and 0.55

0.6 × 0.85 gets 1 mark.

When moving along a path multiply the probabilities on each of the branches.

Consider all the possible ways to catch a bus back home as shown in blue on the probability tree diagram.

WARNING

A common error is to omit one of the possible ways.

b

To school   Back home

Walk   0.85   Walk
0.6   0.15   Bus

Bus   0.45   Walk
0.4   0.55   Bus

$P(\text{Back home by bus}) = P(WB) + P(BB)$
$= 0.6 \times 0.15 + 0.4 \times 0.55$
$= 0.09 + 0.22 = 0.31$

Either 0.6 × 0.15 or 0.4 × 0.55 gets 1 mark.

0.09 + 0.22 gets 1 mark.

WARNING

A common error is to write $\frac{11}{20}$, $\frac{5}{20}$ and $\frac{1}{20}$ instead of $\frac{11}{19}$, $\frac{5}{19}$ and $\frac{1}{19}$; forgetting that only 19 remain after the first sock has been chosen.

Robin has 20 socks in a drawer.
Twelve of the socks are red.
Six of the socks are blue.
Two of the socks are white.
He picks two socks at random from the drawer. Calculate the probability that he chooses two socks of the same colour.

(1385 June 2000)

$P(\text{same colour}) = P(RR) + P(BB) + P(WW)$

$= \frac{12}{20} \times \frac{11}{19} + \frac{6}{20} \times \frac{5}{19} + \frac{2}{20} \times \frac{1}{19}$

$= \frac{132}{380} + \frac{30}{380} + \frac{2}{380}$

$= \frac{164}{380}$

Writing the three products gets 2 marks (just one product gets 1 mark).

Adding the three products gets 1 mark.

EXAM TIP

Further simplification of this fraction is usually not required in a probability question.

United fan, Carol, thinks that the probability that United will reach the second round of the cup is 0.95
She thinks that if they reach the second round, the probability that they will reach the third round is 0.8

a Calculate the probability that they will reach the second round but not the third round.
b Calculate the probability that they will not reach the third round.

a

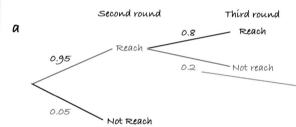

Second round   Third round

0.8 — Reach
0.95 — Reach
0.2 — Not reach
0.05 — Not Reach

$(1 - 0.8)$ gets 1 mark.

Writing 0.95×0.2 gets 1 mark.

This gets 1 mark.

P(Reach, Not reach) = 0.95 × 0.2 = 0.19

$0.05 + 0.19$ gets 1 mark.

b P(Not reach the third round)
= P(Not reach the second round) + P(Reach, Not reach)
= 0.05 + 0.19
= 0.24

Now try these

1 There are 9 stones in a bag.
4 stones are blue.
5 stones are green.
Lisa takes a stone at random from the bag.
She **does not replace it**.
She then takes at random a second stone from the bag.
Work out the probability that at least one of these two stones is blue.

(2544 March 2007)

2 There are 4 red balls, 5 blue balls and 3 green balls in a bag.
A ball is to be taken at random and not replaced.
A second ball is then to be taken at random.

   a   Complete the tree diagram below.

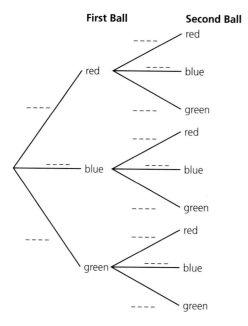

   b Use the tree diagram to calculate the probability that both balls taken will be

      i red,   **ii** the same colour.

   c Calculate the probability that exactly one of the balls taken will be red.

(1384 November 1997)

3 A packet contains stamps from three different countries.
The packet contains 4 Spanish stamps, 10 French stamps and
6 German stamps.
Two stamps are to be removed at random, without replacement.
Calculate the probability that both stamps will be from the same country.

(1385 June 1998)

4 There are 8 balls in a box. 5 balls are red and 3 balls are blue.
Simon takes a ball at random from the box and notes its colour.
He does not replace it.
Simon then takes at random a second ball from the box.

   a Work out the probability that the first ball is red and the second ball is blue.

   b Work out the probability that the balls are of different colours.

5 Kevin has 2 bags of coloured discs.
   Bag **P** contains 4 red discs and 1 white disc.
   Bag **Q** contains 2 red discs and 5 white discs.
   Kevin throws a fair 6-sided dice. If the dice lands on 1, he takes a disc
   at random from bag **P**. If the dice does not land on 1 he takes a disc at
   random from bag **Q**.
   Calculate the probability that Kevin takes a white disc.

6
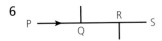

   Becky is driving along a road from P. The probability that she will turn left at
   Q is 0.4
   If she continues along the road at Q, the probability that she will turn right
   at R is 0.75

   a   Calculate the probability that Becky will reach S without turning off
       the road.
   b   Calculate the probability that Becky will not turn right at R.
   c   Calculate the probability that Becky will turn off the road before
       reaching S.

Key terms and exam vocabulary

Instruction	Meaning	Example		Notes
Write down	Little or no working is necessary.	Q The length of a line is 8 cm, correct to the nearest centimetre. Write down the **least** possible length of the line. A 7.5 cm		When "Write down" appears, the question is usually worth 1 mark.
Work out Calculate	These mean the same. Working or calculation is expected.	Q Work out an estimate for $\dfrac{496 \times 6.3}{0.48}$ A Estimate $= \dfrac{500 \times 6}{0.5} = \dfrac{3000}{0.5}$ $= 3000 \times 2 = 6000$		There will usually be marks allocated to working.
Find	It could mean 'Write down' or it could mean 'Work out' or 'Calculate'.	Q Here is a list of numbers. 3 7 8 5 7 a Find the mode. b Find the mean. A a Mode = 7 b Mean $= \dfrac{3 + 7 + 8 + 5 + 7}{5} = \dfrac{30}{5} = 6$		Part **a** is "Write down" and part **b** is "Work out"
Simplify	These mean the same. "Fully" emphasises that the answer must be expressed as simply as possible.	Q Simplify $4x + 5y - x + 3y$ A $3x + 8y$		$3x + 5y + 3y$, for example, is also simpler but would not score full marks.
Simplify fully		Q Simplify fully $(5x^3y^2)^2$ A $25x^6y^4$		Used mainly in Higher tier algebra questions. Incomplete simplification such as $5^2x^6y^4$ would not score full marks.
Factorise	These mean the same. "Completely" warns the candidate against incomplete factorisation.	Q Factorise $x^2 - 8x$ A $x(x - 8)$		Only a single term can be taken outside the brackets.
Factorise completely		Q Factorise completely $6x^2 + 9x$ A $3x(2x + 3)$		Incomplete factorisation such as $3(2x^2 + 3x)$ would not score full marks.
Expand Multiply out	These mean the same.	Q Expand $5(x - 4)$ Q Multiply out $5(x - 4)$	A $5x - 20$	The answer cannot be simplified.
Expand and simplify	Like terms must be collected so that the answer is expressed as simply as possible.	Q Expand and simplify $(x + 4)(x - 1)$ A $x^2 - x + 4x - 4$ $= x^2 + 3x - 4$		$x^2 - x + 4x - 4$ would not score full marks.

Solve	Find the number which the letter in an equation stands for.	Q Solve $3(x + 5) = 12$ A $3x + 15 = 12$ $\quad\quad 3x = 12 - 15$ $\quad\quad 3x = -3$ $\quad\quad\quad x = -1$	$x = -1$ is called the **solution** of the equation. For full marks, $x = -1$ must be stated.
Construct	Use a ruler and compasses only to draw a shape. The ruler may be used to measure lengths, apart from the following four 'straight edge and compasses' constructions • an equilateral triangle with a given base • a hexagon inside a circle • perpendicular bisector of a given line • bisector of a given angle	Q The lengths of the sides of a triangle are 4.3 cm, 3.8 cm and 2.9 cm. Use ruler and compasses to construct this triangle accurately. You must show all construction lines. A 3.8 cm 2.9 cm 4.3 cm Q A point moves so that it is always equidistant from these two fixed lines. Construct its locus. A	Construction lines and arcs must be clearly visible. Don't rub them out. When drawing lines, you are expected to be within 1 mm of the required length.

		Q *ABCD* is a parallelogram. *AB* = 4.2 cm, *BC* = 3.9 cm and angle *ABC* = 141°. **a** Make a sketch of the parallelogram *ABCD*. **b** Make an accurate drawing of the parallelogram *ABCD*.	
Make a sketch	In a sketch, accurate lengths and angles are not required but it must be clear and as realistic as possible.		Use a ruler to draw the straight lines in a sketch. Notice that angle *ABC* is drawn as an obtuse angle.
		A a 	
Make an accurate drawing	Use a ruler, compasses and a protractor to draw a shape accurately.	**b** 	Construction lines should be visible. When drawing lines, you are expected to be within 1 mm of the required length. When drawing angles, you are expected to be within 2° of the required size.

Answers

1 Reverse percentages

1. £130
2. 1800
3. £170
4. £340
5. £650
6. £275
7. £151
8. £275
9. £6500
10. £2720

2 Standard form

1. a 4.75×10^7 b 6×10^{-5}
2. a 5×10^7 b 8.2×10^{-5}
3. a 3800   b 4.5×10^{-4}
4. a 6.52×10^4 b 0.0836
5. a 3.2×10^{-2} b 15 800
6. a 7.68×10^7 b 3.5×10^{-4}
7. a 4.56×10^5 b 3.4×10^{-4}
   c 1.6×10^8
8. a i 50 100   ii 9×10^{-4}
   b 2.4×10^9
9. a i 4×10^7 ii 0.000 03
   b 1.2×10^3
10. 2.83×10^{15}
11. 8.01×10^{10}
12. a 5.72×10^6 b 1.4×10^{-7}
13. 5.8×10^{-4}
14. 1.7×10^{12}
15. 1.12×10^2
16. a 1×10^{-9} b 2×10^8
17. a 4.2×10^5 b 2.4×10^{-6} grams

3 Bounds

1. a 46.75   b 14
   c 4   d 1.89 (3 s.f.)
2. a 97.75   b 20
   c 2   d 1.21 (3 s.f.)
3. a $223.25\,\text{m}^2$ b $257.25\,\text{m}^2$
4. $n = 1.495$, $r = 0.45$
5. a $29.25\,\text{cm}^2$ b 0.538 cm (3 s.f.)
6. a upper bound of $r = 1.75$; lower bound of $r = 1.65$;
   upper bound of $R = 31.05$; lower bound of $R = 30.95$
   b 29.2   c 5531
7. upper bound = 24.020 km/h; lower bound = 23.980 km/h
8. a i 6.75 cm   ii 6.65 cm
   b i 4.05263… cm   ii 3.977777… cm
9. 75.879
10. i 0.888 (3 s.f.)   i 0.924 (3 s.f.)

4 Surds

1. a 2   b 5
   c 2   d 10
   e 9   f 3
2. a $7 + 3\sqrt{3}$ b $5 - 6\sqrt{2}$
   c $14 + 6\sqrt{5}$ d $5 - 2\sqrt{3}$
3. a $\dfrac{\sqrt{3}}{3}$ b $\dfrac{3\sqrt{5}}{5}$
   c $2\sqrt{2}$ d $3\sqrt{5}$
4. a $1 + 2\sqrt{3}$ b $2 + 5\sqrt{3}$
5. a 14 cm   b $(7 + 3\sqrt{3})\,\text{cm}^2$
6. $2 + 3\sqrt{2}$
7. $-2 + 3\sqrt{3}$
8. a $2\sqrt{2}$ b $5\sqrt{2}$
   c $\dfrac{2 + \sqrt{2}}{2}$
9. a $\dfrac{3}{2}$ b $16\sqrt{2}$
   c $\dfrac{\sqrt{2}}{32}$
10. a $\dfrac{\sqrt{7}}{7}$ b i $18 + 6\sqrt{5}$ ii 2

5 Solving equations that have fractions

1. $\dfrac{2}{5}$
2. 5.5
3. 7
4. −0.8
5. 0.75
6. 10.5
7. 0.8
8. a $x = 2.25$
   b $y = -0.5$, $y = 2.5$
9. −1, 8
10. −3.56 (2 d.p.), 0.56 (2 d.p.)

6 Changing the subject of a formula

1. $v = \dfrac{I + mu}{m}$
2. $u = \sqrt{v^2 - 2as}$
3. $W = Ih^2$
4. $p = \sqrt{\dfrac{2a - M}{2}}$ or $\sqrt{a - \dfrac{M}{2}}$
5. $b = \sqrt{\dfrac{3V - \pi a^2}{\pi}}$ or $\sqrt{\dfrac{3V}{\pi} - a^2}$
6. $g = \dfrac{4\pi^2 l}{T^2}$

7 $v = \dfrac{fu}{u - f}$

8 $a = \dfrac{n^2 - Pn}{P - 1}$

9 $x = \sqrt{\dfrac{k}{y}} - a$

10 $g = \dfrac{pc}{(q - p)}$

7 Using $y = mx + c$

1 $y = 3x + 2$, $y = 3x - 3$

2 $y = -2x + 3$, $2y = x + 4$

3 $y = -x + 3$

4 **a** 2

  b $y = 2x - 1$

  c For example, $y = 2x + 3$ (any line with $m = 2$)

5 $3y = -2x + 7$

6 **a** For example, $y = \frac{1}{2}x + 5$ (any line with $m = \frac{1}{2}$)

  b For example, $y = 3x + 1$ (any line with $c = 1$)

  c $y = -2x + 26$

7 **a** $y = -\frac{2}{3}x + \frac{4}{3}$

  b $y = 1.5x - 3$

8 Solving inequalities graphically

1

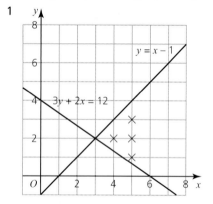

2 i C   **ii** B

3

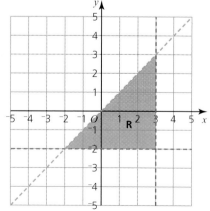

4

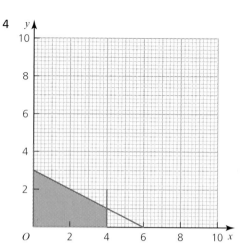

5

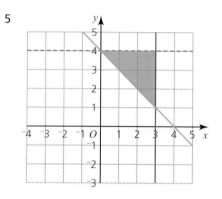

6

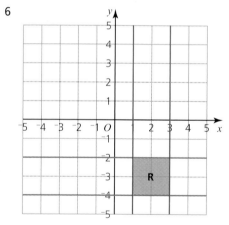

9 Direct proportion and inverse proportion

1 **a** $y = 1.25x$ **b** 16.25

2 **a** $y = 0.8x^2$ **b** 28.8   **c** 3.5

3 **a** $y = 4\sqrt{x}$ **b** 16

4 **a** $y = \dfrac{96}{x}$ **b** 48   **c** 38.4

5 **a** $y = \dfrac{40}{x^2}$ **b** 1.41

6 **a** $y = \dfrac{60}{\sqrt{x}}$ **b** 20   **c** 36

7 **a** $D = \dfrac{3}{160}t^2$ **b** 76.8   **c** 25.3

8 a 960   b 560
9 a $d = 5t^2$ b 45 m
   c 11 seconds
10 a $n = \dfrac{30}{\sqrt{l}}$ b 50
   c 2.25 m

10 Further factorising of quadratic expressions

1 a $(y + 3)(y - 3)$ b $(11t + 1)(11t - 1)$
   c $(p + 4q)(p - 4q)$ d $(5 + a)(5 - a)$
2 a $(2x + 1)(x + 3)$ b $(5x - 3)(x - 1)$
   c $(5x + 2)(x - 1)$ d $(3x + 4)(x - 2)$
3 $2(x + 5y)(x - 5y)$
4 a $(m + n)(m - n)$ b 100 000
5 a $(3x - 1)(2x + 3)$ b $(3x - 1)^2$
   c $(3x - 1)$
6 $(2x - 3)$
8 a $(x - 2)(2x - 3)$ b i $2a(n - a)$

11 Solving quadratic equations using the formula

1 −0.586, −3.41
2 −0.697, −4.30
3 −0.293, −1.71
4 6.27, −1.27
5 2.39, 0.279
6 7.41, −0.405
7 0.414, −2.41
8 0.464, −6.46
9 4.45, −0.449
10 1.21, −0.207
11 2.79, −1.79
12 2.00, −0.500
13 4.32, −2.32
14 1.83, −1.83

12 Solving problems using quadratic equations

1 a $x = -9$ or $x = 5$ b 9
2 a $x = -5$ or $x = 3$ b 7.5 cm²
3 a $x = -11$ or $x = -3$ b 3 cm and 5 cm
4 a $x = -6$ or $x = 4$ b 3 hours
5 a $x = 4$ or $x = -2$ b 2 cm, 2 cm, 8 cm
6 a $x = -9$ or $x = 10$ b 10 CDs
7 a $x = 2.171$ or $x = -3.838$
   b 4.34 cm
8 a $x = 1.424$ or $x = -8.424$
   b 13.7 cm

13 Completing the square

1 $(x + 4)^2 + 1$
2 $(x + 2)^2 - 4$
3 $(x - 4)^2 - 15$
4 $\left(x - \dfrac{1}{2}\right)^2 + \dfrac{3}{4}$
5 a i $p = 3$ ii $q = 5$
   b (3, 5)

6 $p = -2, q = 11$
7 a $p = 3, q = -9$ b $x = -3 \pm \sqrt{5}$
8 $a = 49, b = -7$
9 $(x - 3)^2 - 6, -6$
10 $45 - (5 + x)^2$, 45 at $x = -5$

14 Simplifying algebraic fractions

1 $x - 1$
2 $\dfrac{7}{2x}$
3 $\dfrac{3}{x - 2}$
4 $\dfrac{6 - x}{6}$
5 $\dfrac{4 + 5x}{x(x + 3)}$
6 $\dfrac{x + 3}{x}$
7 $\dfrac{y}{y - 2}$
8 $\dfrac{x + 3}{x - 6}$
9 $\dfrac{2x + 5}{x + 4}$
10 $\dfrac{4n}{(2n - 1)(2n + 1)}$
11 $\dfrac{-2x - 7}{(x + 2)(x - 1)}$
12 $\dfrac{3}{x + 4}$

15 Solving simultaneous linear and quadratic equations

1 $x = \dfrac{1}{2}, y = \dfrac{3}{2}$ or $x = 1, y = 3$
2 $x = -1, y = 4$ or $x = 4, y = 4$
3 (3, 3) (−2, −2)
4 $\left(-\dfrac{1}{4}, \dfrac{1}{4}\right)$ and (1, 4)
5

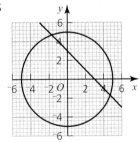

From the graph $x = 4.7, y = -1.7$ or $x = -1.7, y = 4.7$
6 $x = 1, y = -2$ or $x = 2, y = -1$
7 $x = -1, y = -5$ or $x = 5, y = 1$
8 6
9 $x = -3, y = -3$ or $x = \dfrac{3}{5}, y = 4\dfrac{1}{5}$
10 $x = 1.8, y = 4.7$ or $x = -2.6, y = -4.3$

16 Algebraic proofs

3 For example, $n = 4$ gives $2n + 1 = 9$ which is a square number.

4 For example, $2 + 3 = 5$ which is prime.

8 a $(p - q)(p + q)$ **b** $n^2 - 1$

17 Similar triangles

1 a 4.8 cm   **b** 2.4 cm
2 a 39 cm   **b** 30 cm
3 a 3.5 cm   **b** 2.7 cm
4 a 24 m
5 a 8 cm   **b** 19.8 cm
6 a 12 cm   **b** 9 cm
7 a 11.6 cm   **b** 6.3 cm
8 a 8 cm   **b** 5.7 cm

18 Area and volume of similar shapes

1 90 cm^2
2 a $x = 20$ **b** $y = 6$
3 a $x = 20$ **b** $y = 10$
4 a 96 cm^2 **b** 750 cm^2
5 39.0625 cm^3
6 3800 cm^2
7 a 300 cm^3 **b** 26.1 m^2
8 a 254.47 m^3 **b** 195 tonnes
   c 960 cm^3

19 Arc length and sector area

1 a i 4.89 cm   **ii** 12.6 cm
    iii 9.60 cm   **iv** 3.40 cm
    v 8.17 cm   **vi** 17.6 cm
  b i 17.1 cm^2 **ii** 56.5 cm^2
    iii 24.0 cm^2 **iv** 6.64 cm^2
    v 24.5 cm^2 **vi** 72.2 cm^2
2 a 12.6 cm   **b** 240 cm^2
3 a 38.2 cm^2 **b** 474 cm^2
4 a 21.8 cm   **b** 66.4 cm^2
5 39.4 cm^2
6 75 cm^2

20 Pythagoras' theorem and trigonometry in 2-D

1 177 cm^2
2 a 7.55 cm   **b** 32.2°
3 31.7 cm^2
5 a 62.0°   **b** 19.1m
6 a 622 km   **b** 600 km
7 a 16.6 m   **b** 30.4°
8 496
9 25.3 m
10 35°

21 Pythagoras' theorem and trigonometry in 3-D

1 a 8.54 cm   **b** 5.83 cm
   c 9.90 cm

2 a 15.6 cm   **b** 30.8°
3 longest length of pencil that would fit inside the cylinder is 12.8 cm
4 a 13 cm   **b** 67.4°
5 a i 11.2 cm   **ii** 16.4 cm
    iii 15.6 cm
  b 47°
6 a i 9.01 cm   **ii** 21.9 cm
  b 21.4 cm
  c 69.4°
7 a 17.2 m   **b** 59.4 m

22 Volume and surface area of 3-D shapes

1 137 000 cm^3
2 a 576 cm^3 **b** 361 cm^2
3 a 39.0 m^2 **b** 42.7 m^3
4 15 cm
5 382 cm^3
6 65π cm^2
7 6.05 m^3
8 10.6 cm
9 15π cm^3
10 a $(2x - 3)(x + 11)$ **b** $\pi \times 3 \times 5 + 2 \times \pi \times 3^2$
   c 11 cm

23 Using the sine rule, the cosine rule and $\frac{1}{2}ab \sin C$

1 53.8 cm^2
2 10.4 cm
3 a 6.43 cm   **b** 11.3 cm^2
4 a 105.7°   **b** 10.7 cm^2
5 100 m
6 a 11.1 cm   **b** 29.8°
7 23.9 cm^2
8 a 23.9 cm^2 **b** 78.4°
9 18.9 m^2
10 a 10.6 km   **b** 1.41 km

24 Circle geometry

1 a 90° (angle in a semicircle is a right angle)
  b 35° (angle at centre = 2 × angle at circumference) (alternatively, sum of angles on a straight line and isosceles triangle can be used)
2 a 70°
  b 20° (angles in the same segment)
3 a i 61°
    ii The sum of the opposite angles of a cyclic quadrilateral is 180°.
  b 29°
4 a i 42°
    ii Angle at centre = 2 × angle at circumference
  b 132°
5 a 40°
  b 12°
  c because angle $EDA \neq 90°$ (angle in a semi-circle = 90°)

6 a 132° (alternate angles; the sum of the opposite angles of a cyclic quadrilateral is 180°)

   b 24° (angle sum of a triangle = 180°; isosceles triangle with equal angles opposite equal sides; alternate segment theorem)

25 Using vectors to solve 2-D geometric problems

1 a i $p + q$ ii $q - p$

   b $\frac{1}{2}(p + q)$

2 $\frac{2}{3}b + \frac{1}{3}a$

3 a i $a + \frac{1}{2}b$ ii $b + \frac{1}{2}a$

   iii $b - a$

   b $\overrightarrow{XY} = \overrightarrow{XP} + \overrightarrow{PY} = -(a + \frac{1}{2}b) + (b + \frac{1}{2}a)$

   $= -\frac{1}{2}a + \frac{1}{2}b = \frac{1}{2}(b - a) = \frac{1}{2}\overrightarrow{QS}$

   so XY is parallel to QS and $XY = \frac{1}{2}QS$

4 a i $b - a$ ii $2a$

   iii $2b - 2a$

   b MN and QR are parallel, the length of QR is twice the length of MN.

5 a i $b - 3a$ ii $2b - 2a$

   iii $\frac{1}{2}b - \frac{1}{2}a$

   b $\overrightarrow{PR} = 4\overrightarrow{PQ}$ so PR and PQ are parallel with the point P in common hence PQR is a straight line

   c 12 cm

6 $\frac{3}{2}(b - a)$

7 a i $-a + b$ ii $2b + 2c$

   iii $2b + 2c - 2a$ iv $b - a$

   b Equal and parallel

8 a i $a + b$ ii $b - a$

   b i $-\frac{1}{3}(a + b)$ iii $\frac{2}{3}a - \frac{1}{3}b$

   iii $\frac{4}{3}a - \frac{2}{3}b$

   c i $\overrightarrow{TR} = 2\overrightarrow{PT}$ so PT and TR are parallel with T in common so PTR is a straight line

   ii $\frac{PT}{PR} = \frac{1}{3}$

9 a i $a + b$ ii $2a - b$

   b $\frac{2}{3}a - \frac{1}{3}b$

   c $\overrightarrow{ST} = 3\overrightarrow{SX}$ so ST and SX are parallel with S in common. Hence SXT is a straight line so that X lies on ST and X divides ST in the ratio 1 : 2

10 a $k = 10$

   b i $\begin{pmatrix} 9 \\ 0 \end{pmatrix}$ ii $\begin{pmatrix} 3 \\ -10 \end{pmatrix}$

   c $\overrightarrow{BD} = 3\overrightarrow{BT}$ hence BD and BT are parallel with B in common so that BTD is a straight line. So T lies on BD.

26 Histograms

1

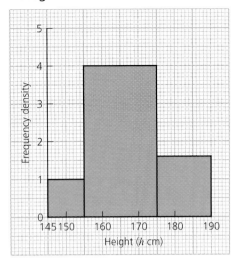

2

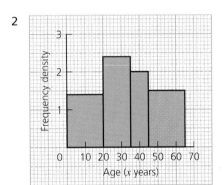

3

Weight (w kg)	Frequency
$2 \leqslant w < 4$	20
$4 \leqslant w < 7$	90
$7 \leqslant w < 9$	50
$9 \leqslant w < 10$	40
$10 \leqslant w < 14$	20

4 a 30

   b $2.2 \times 10 + 1.6 \times 45 = 94$

5

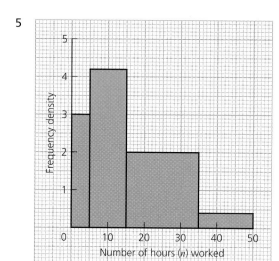

6

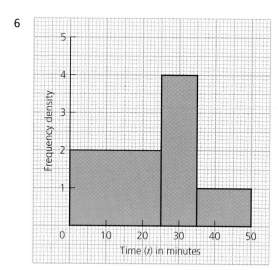

7

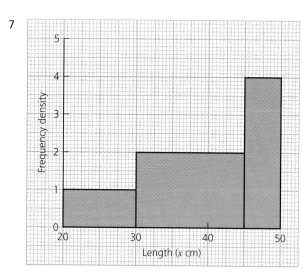

1 i 23 ii 8

2 7

3

School	A	B	C	D	Total
Number of girls	22	14	35	9	80

4 A 12, B 20, C 18

5 **a** 6

b alternative A: Number of Year 10 boys in the sample is 17 and number of Year 11 girls in the sample is 6 so Trevor is not correct

alternative B: the numbers in the stratified sample will reflect the relative numbers in the population so Trevor is correct. The actual numbers (17 and 6) reflect rounding as you cannot have part of a person

6 30

7 Greek 7, Spanish 19, German 15, French 29

8 **a** 8 **b** 85 or 86

9 **a** 13

b i each member of the population being sampled is equally likely to be selected

ii pulling names out of a hat, use of random numbers

10 **a** When a sample is to be taken from a population, it might be possible to split the population into a number of groups, called strata.

In a stratified sample, the fraction of the size of each stratum in the sample to the size of the sample is the same as the fraction of the size of the stratum in the population to the size of the population.

b 12

28 Using P(A and B) = P(A) × P(B)

1 0.675

2 0.09

3 **a**

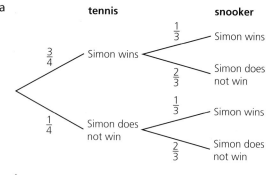

b $\frac{1}{4}$

4 0.0459

5 **i** 0.0026 **ii** 0.8976

6 $\frac{1}{6}$

29 Using P(A or B) = P(A) + P(B)

1 0.3

2 a

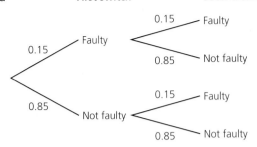

	First switch	Second switch

- 0.15 → Faulty — 0.15 → Faulty
- Faulty — 0.85 → Not faulty
- 0.15 — 0.85 → Not faulty — 0.15 → Faulty
- 0.85 → Not faulty — 0.85 → Not faulty

   b 0.2775

3 $\frac{13}{36}$

4 0.3941

5 a 0.9   **b** 0.1

   c 0.48

30 Conditional probability

1 $\frac{52}{72}$ (or equivalent)

2 a

	First Ball	Second Ball

- $\frac{4}{12}$ → red
  - $\frac{3}{11}$ → red
  - $\frac{5}{11}$ → blue
  - $\frac{3}{11}$ → green
- $\frac{5}{12}$ → blue
  - $\frac{4}{11}$ → red
  - $\frac{4}{11}$ → blue
  - $\frac{3}{11}$ → green
- $\frac{3}{12}$ → green
  - $\frac{4}{11}$ → red
  - $\frac{5}{11}$ → blue
  - $\frac{2}{11}$ → green

   b i $\frac{12}{132}$ (or equivalent)   **ii** $\frac{38}{132}$ (or equivalent)

   c $\frac{64}{132}$ (or equivalent)

3 $\frac{132}{380}$ (or equivalent)

4 a $\frac{15}{56}$

   b $\frac{30}{56}$ (or equivalent)

5 $\frac{132}{201}$ (or equivalent)

6 a 0.15   **b** 0.55

   c 0.85

Index

Published by: Edexcel Limited, One90 High Holborn, London WC1V 7BH

Distributed by: Pearson Education Limited, Edinburgh Gate, Harlow, Essex CM20 2JE, England
www.longman.co.uk

First published 2008
Fourth impression 2010
ISBN 978-1-84690-188-1

Typeset by Pantek Arts Ltd
Printed in China
SWTC/04

The publisher's policy is to use paper manufactured from sustainable forests.